Cyber and Electromagnetic Threats in Modern Relay Protection

Cyber and Electromagnetic Threats in Modern Relay Protection

Vladimir Gurevich

ISRAEL ELECTRIC CORPORATION, HAIFA

CRC Press
Taylor & Francis Group
Boca Raton London New York

CRC Press is an imprint of the
Taylor & Francis Group, an **informa** business

CRC Press
Taylor & Francis Group
6000 Broken Sound Parkway NW, Suite 300
Boca Raton, FL 33487-2742

First issued in paperback 2017

© 2015 by Taylor & Francis Group, LLC
CRC Press is an imprint of Taylor & Francis Group, an Informa business

No claim to original U.S. Government works

ISBN-13: 978-1-4822-6431-9 (hbk)
ISBN-13: 978-1-138-89282-8 (pbk)

Visit the Taylor & Francis Web site at
http://www.taylorandfrancis.com

and the CRC Press Web site at
http://www.crcpress.com

Contents

Preface

Future conflicts will be won in a new arena—
that of the electromagnetic spectrum and cyberspace.
We must merge, then master those realms.

Admiral Jonathan W. Greenert
U.S. Navy

Our power grid is very vulnerable.
It's very much on edge. Our military knows that.

Ex-Congressmen Roscoe Bartlett

The problem is not the technology. We know how to protect against it.
It's not the money, it doesn't cost that much.
The problem is the politics.
It always seems to be the politics that gets in the way.

Peter Vincent Pry, PhD
Executive Director of the Task Force
on National and Homeland Security
President of EMPACT America

It would be "suicidally optimistic" to assume that an EMP attack that inflicted
a state-wide blackout would not also cause cascading grid and infrastructure
failures at least regionally.

Dr. William Radasky and Dr. Peter Vincent Pry

Relay protection occupies a special place in the system of generation, transmission, and distribution of electric energy. It does not take part in production, transfer, or distribution of electric energy directly. In fact, it does not show itself under normal conditions of operation of a power system. If you disconnect it, nothing will change, that is, generators at power plants will continue producing electric energy and power transmission lines, and distribution networks will continue delivering energy to consumers. But this situation is very deceptive: the smallest technical breakdown of equipment can result in the collapse of the electric power system of an entire country if relay protection fails to interfere into this situation. These facts are well known by specialists and do not require additional clarifications. But it appears that everything is not this simple. Modern protection relays consist of sophisticated electronic complexes, which can also fail like any other type of modern electronic equipment. What happens in case relay protection fails while in an emergency mode in electric power systems? Nothing significant happens, since the protection relay is not operating all by itself, but together with several other relay protection devices. If one relay fails to activate, another will step in. After all, all critical power assets have backup protection. But failure to activate is not the only nonoperation of protection relays in the emergency mode. It can be falsely actuated under normal mode of operation as well. This is where the problems begin. The fact is that unnecessary actuation of a relay cannot be *corrected* by backup protection relays. What does unnecessary actuation of protection relay mean? It means the disconnection in power transmitting lines, transformers, and generators by means of switches of thousands of consumers. By no means can the systems of automatic re-closing or automatic takeover always correct the situation. The transient state in electric power circuits and the power system as a whole, which take place during sudden disconnection of high-power units, can result in subsequent disconnections of power transmission lines and generators; in other words, it leads to total outage and collapse of the energy system. The world's majority energy system accidents exhibited this scenario. It appears that protection relays can trigger the collapse of a normally functioning system also.

Recently, people involved in planning potential military campaigns have become aware of this fact. Modern scenarios of power struggles between countries are rarely based on using traditional means of striking lives and weapons of the enemy; they rather rely more and more on means that can affect the enemy's infrastructure but avoid human losses. Damaging the infrastructure of modern postindustrial society proves to be more detrimental than ordinary military actions. Electronification and the dependence of any developed country's infrastructure on computers make destruction of the infrastructure significantly simpler, since the destruction can be virtual rather than physical. Thus, the more developed the infrastructure is, the more vulnerable it will be to virtual impact.

> Some foreign analysts, judging from open source statements and writings, appear to regard EMP attack as a legitimate use of nuclear weapons because EMP would inflict no or few prompt civilian casualties. EMP attack appears to be a unique exception to the general stigma attached to nuclear employment by most of the international community in public statements (Report: "Terrorism and EMP Threat to Homeland Security" – Subcommittee on Terrorism Technology and Homeland Security, S/n J-109-5, March 8, 2005)

What place does relay protection occupy in the infrastructure of a country? An absolutely special place, since through protection relays, which control the position of circuit breakers, one can gain access to change the configuration of a power electric network remotely, which results in the collapse of an ordinary functioning power system. Today, this is clear to organizations strategizing battles. Dozens of large corporations from all over the world are working on orders to create special types of equipment, which can affect very sensitive electronic equipment of the modern power industry. Digital protection relays, due to their special position, are by far not the last target to be hit in the first round. Today, two types of remote destructive impacts on digital systems are known: cyber attacks and intentional remote destructive electromagnetic impacts.

Modern trends of relay protection development include, but are not limited to, the overall transition to digital relays, the continuous sophistication of their software and hardware, the increase in the number of functions that they perform (including those that are not directly related to relay protection), the transition from fiber-glass communication channels to less protected channels (Ethernet, Wi-Fi), the continuous miniaturization of electronic equipment, the use of flash-memories based on changing and registration of a very weak electric charge in the insulated area of transistor that is getting wider, and the increase in the number of transistors in microprocessors and the reduction of their operating voltage that make remote destructive impacts significantly easier. On the one hand, we see a continuous increase in relay protection susceptibility; on the other hand, we see a continuous improvement in the methods of remote destructive impact. As a result, these two dangerous vectors of development are rapidly heading toward each other. To recall the famous saying of Winston Churchill, "The Stone Age may return on the gleaming wings of science."

The situation is worsening because both criminals and terrorist organizations are gaining access to modern means of impact on computer and microprocessor systems. This makes the meeting of these two vectors inevitable. This is why it is necessary to understand the existing danger and take preventive measures in advance.

In this book, the author attempts to convince the reader in actuality of this danger and presents solutions to the problem.

Please send your remarks about the book to the author: vladimir.gurevich@gmail.com

Dixi et animam meam salvavi!

Abstract

The book provides a detailed overview of the vulnerabilities of digital protection relays (DRP) to natural and intentional destructive impacts, which include cyber attacks and electromagnetic impacts. Modern technical tools that realize intentional remote destructive impacts to DPR are also described. The book discusses both traditional passive means of protection, such as screened cabinets, filters, cables, special materials and covers, and advanced tools based on circuit and hardware methods.

The book is intended for engineers dealing with the development, designing, and use of relay protection and can be beneficial for scientists, teachers, postgraduates, and students of specific subjects in vocational schools and higher education establishments.

Author

Vladimir I. Gurevich received an MS in electrical engineering (1978) at the Kharkov Technical University and a PhD (1986) from the Kharkov National Polytechnic University, Kharkov, Ukraine.

Throughout his employment experience, he has been in the following positions: teacher, assistant professor, and associate professor at Kharkov Technical University, and chief engineer and director of Inventor, Ltd.

In 1994, he arrived in Israel and works today at Israel Electric Corp. as a senior specialist and head of section of the Central Electric Laboratory, Haifa.

He is the author of more than 180 professional papers and 11 books and holder of nearly 120 patents in the field of electrical engineering and power electronics. In 2006, he was honorable professor with the Kharkov Technical University.

Other books of the author published by Taylor & Francis Group:

- *Protection Devices and Systems for High Voltage Applications*
- *Electrical Relays: Principles and Applications*
- *Electronic Devices on Discrete Components for Industrial and Power Engineering*
- *Digital Protective Relays: Problems and Solutions*
- *Power Supply Devices and Systems of Relay Protection*

1

Technological Advance in Relay Protection: Dangerous Tendencies

1.1 Issues of Philosophy in Relay Protection

Philosophy beginning with the ancient Greeks is literally "a love of wisdom" and deals with the most general issues of reality. The Wikipedia terms philosophy as a science which studies everything. Logic and critical analysis are the pillars of philosophic thinking. So why don't we use these attributes of philosophy to analyze the situation around relay protection (RP)? It appears that such analysis may result in counterintuitive findings (Figure 1.1).

The life of a modern person is closely related to the use of complex and interconnected systems such as cellular communication, television (TV), radio, and electric systems. All these systems can be visualized as a so-called "consumer chain" that consists of series of links. The last link in this chain, that is, the one that directly interacts with a person, would be a certain apparatus (terminal): cellular phone, TV, radio receiver, refrigerator, washing machine, etc. The aspiration to improve the last link (in other words, the one that actually interacts with a person) and make it perfect, even though all other links may be far away from perfection, is clear and justified. A special design of a TV set, its user-friendliness, and special functions (such as record and playback telecasts according to a specific schedule, playback CDs, and split screen that allows having a main screen and a series of auxiliary screens, making it possible to watch several channels simultaneously) add significant value to such a TV set from the standpoint of the consumer, regardless the fact that this TV set is only a final link in a long chain called television. It does not mean that the quality of TV programs or the quality of broadcasting will be of the same quality (perfect) as the final link. However, this doesn't prevent a rich consumer from investing in an expensive (perfect) final link. Likewise, poorer consumers do not stop dreaming about this perfect final link. Thus, the final link in different consumer chains has a special status, and certain requirements and attention are accorded to it by both consumers and manufacturers. On the other hand, regardless of

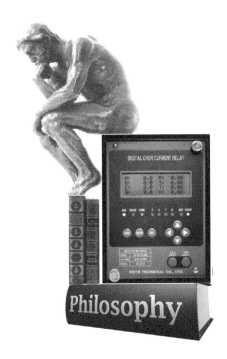

FIGURE 1.1
Philosophy in RP.

the perfection of the final link, it cannot influence the quality or reliability of the chain in general. Indeed, a broken TV set in one of the rooms in the consumer's house will not influence the operation of TV sets in other rooms or the neighbors' TV sets.

Another feature of the final link of the aforementioned consumer chains is the applying of customer requirements to functionality and design beyond the requirements of reliability and longevity. This is conditioned by modern trends, when substitution of one final link by the other has not much to do with malfunctioning or breakage, but with technological obsolescence and the emergence on the market of new models with better functions and improved design.

Now, let us compare this situation with what happens in RP, which is the most important component of a consumer power supply circuit that consists of a series of links called production, transmission, and distribution of electric energy. Where is the place of RP in this circuit? Surprisingly, there is no such link in this chain! Indeed, RP neither participates nor influences the operation of the circuit under normal mode of consumer chain operation. RP does not influence the amount of produced energy. Nor does it influence the capacity of energy transmitting lines or the process of energy distribution. RP can even be disconnected from energy supply circuit, and there will be no effect on the circuit's operation. So what is RP and where is its place in

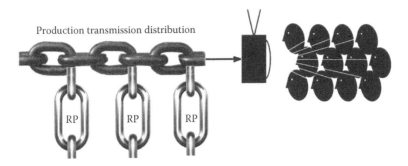

Production transmission distribution

FIGURE 1.2
Visualization of a consumer chain of power supply equipped with RP.

energy transmission and distribution circuit? Visually RP can be depicted in this chain as a set of separate auxiliary links installed in the places of connection of main links of the energy supply circuit, that is, production, transmission, and distribution of electric energy (Figure 1.2).

Functionally, these places of connection are formed by high-voltage (HV) switches, the condition of which is determined by RP. In other words, even though RP is not a series link in the power supply circuit, it can influence the connectors between the links (by circuit breakers) by cutting the ties between all the links of this chain. This is a fundamental difference of RP from other links in consumer chains.

If RP does not influence the power supply circuit under its normal mode of operation, does it have any effect under the emergency mode? It is widely perceived that this influence consists in the prevention of emergency modes in the power supply circuit. Is this really so? To answer this question, we need to understand what RP is and what its functions are. Let us review publication [1], which provides a detailed analysis of this issue based on which we obtain such concepts as "protective relay" and "RP":

> Protective relay is a device, the purpose of which is to detect the emergency mode of the object being protected and send a command to a power control element, which eliminates this mode.
> A system of relay protection is an aggregate of related devices, which ensure detect of the emergency mode in the operation of electric equipment and its elimination.

These definitions show that regardless of the widespread opinion that an RP cannot prevent the emergency mode in a power supply circuit, it can limit the scale of its effect on this circuit in time and space, in other words, limit the material damage from the breakdown and nothing more.

Everything said earlier is related to a properly functioning RP and its correct operation. But as many other complex technical devices, RP can also malfunction. This creates an absolutely different situation, where a malfunctioning RP due to a so-called "unnecessary protection operation" can send a faulty command to open a circuit breaker (in other words, break ties between the links of a power supply circuit), thus creating an artificial prevention of normal functioning of a power supply circuit, that is, its switching to an abnormal, emergency mode, leading to the disconnection of thousands of consumers and great damage.

This makes us conclude that RP cannot prevent an emergency mode of operation of a power supply circuit, but it can cause this mode.

Recently, significant qualitative changes have happened in the field of RP. Single-function electromechanical protection relays (EMRs) have been replaced by multifunctional digital protective relays (DPRs) with much higher qualitative characteristics and easily programmable logic. How does perfection of characteristics and improved functional capabilities of the new protection relays influence the operation of power supply circuit? As mentioned earlier, there is no influence under the normal mode of operation. However, in the case of emergency mode in the circuit, the DPR can efficiently limit its effect in time and space due to their improved characteristics; in other words, they are more effective in limiting the material damage than EMRs.

At the same time, it is known [2] that DPRs are less reliable than EMRs (we are talking about the best electromechanical relays manufactured by the leading Western companies). Their lifetime does not exceed 15–20 years. They are more susceptible to destructive external impacts, such as cyber attacks or intentional electromagnetic impact. A lot of functions in one terminal, some additional functions of DPR not specific to RP, and mistakes of staff during free programming of logics reduce reliability of RP even further and increase the probability of malfunctioning, that is, resulting in such impacts on the power supply network that lead to deterioration of its operation (emergency modes).

Thus, transition from EMRs to DPR results in reduction of equipment damage from accidental emergencies in the power supply network, but at the same time, it leads to an increase in the number of accidents (due to additional accidents caused by the malfunctioning of RP device itself) in the power system. To support this, let us quote [3], which provides a general picture of the situation:

> Conventional electro-mechanic relay protection, like all domestic low-voltage equipment is reliable and long-lasting, which corresponds to the main principle of a renowned Austrian company of Paul Hertz: "All types of electro-technical equipment must operate more than 50 years." It is noteworthy that the unique power plant of Russia has been working without system outages for 50 years. In the course of reformation of electric power sector the philosophy of relay protection and automatics (RPA), where mostly electro-mechanical and micro-electronic devices produced by Cheboksary Electric Apparatus Plant

have been used, is reviewed now. Technical re-equipment of protection presupposes implementation of microprocessor RPA, produced mainly by foreign companies. This decision rests on the positive experience of implementation of microprocessor-based devices abroad. But it should be remembered that user friendly and multi-functional foreign equipment has its specific features. According to International power organization, which dates back to the time of full-fledged socialism (SIGRE) and the Soviet scientist Venikov V.A., system outages are regular in the power industry of many countries where equipment of companies sharing the Russian market has been used and is used now. Unlike the trouble-free operation of the Unified electric power system of Russia, which is protected by electro-mechanical relays, there were 13 major accidents abroad during the last two decades and 8 of them were in the USA. Power supply failures covered large territories, whereas Russian equipment continues functioning faultlessly in Egypt, Iran and Africa. There were neither breakdowns nor failures at power plants.

Taking this into consideration, we come to the conclusion that unlike consumer chains mentioned in the beginning of this chapter, reliability and longevity of RP in the power supply circuits should prevail beyond improved characteristics and expanded capacity and design.

1.2 Extrusion into the Historical Domain

For over a hundred years, all the tasks of RP have been performed by EMRs. The fact that EMRs are still widely used in many countries, including Russia (about 80% of all types of protective relays), proves that in general EMRs are capable of solving all the present problems of the RP. However, during the past 15–20 years, there has been a widespread displacement of EMR by DPR. DPR and various programmable logic controllers (PLCs) that control the operating modes of electrical equipment have become an integral part of our lives, and in many cases, there is no other device available to substitute for them to ensure the normal functioning of the power industry. This is not due to some unique features of microprocessor devices, this is rather a result of the costs of the fully automated production of DPR based on printed circuit boards compared with the production of high-precision mechanics for the relays of the previous generation. Thirty to forty years ago due to the necessity for cutting production costs and improving profitability of production, the development of the new types of EMRs was stopped and all efforts were focused first on the creation of the static solid-state protections and then on the development of DPRs. The first types of DPR simply copied all the functions and characteristics of the relays of previous generations.

New features and capabilities of DPR have been implemented only many years after. This technical policy of manufacturers has resulted in the complete halting of the production of all other types of protection by all of the world's leading manufacturers of the RP, and DPRs have become nearly the only available type of protection.

Even the very first DPRs, which simply copied the functions of the static solid-state transistor-based relays (see Figure 1.3), revealed serious problems of the DPRs: more frequent failures and irreparability due to the presence of the special microprocessor and nonvolatile memory containing the program. As a result, while the relays of type RXIDF-2H built on transistors and other discrete components were quick to repair and return to operation, their microprocessor-based analog RXIDK-2H must be discarded. Hence, the microprocessor-based RXIDK-2H has long been taken out of service while RXIDF-2H is still used. The tendency of the RP reliability weakening associated with the transition to the DPR and noticed at the beginning of this process can be traced so far, despite the fact that modern generation of the DPRs has little in common with the first samples manufactured a few decades ago (see Figure 1.1). This goes to prove that the problem is systemic rather than a result of the single technical defects specific to early DPR models. But no one wanted to gain the character of retrograde and nobody wanted to talk about the obvious problems of DPRs welcomed with such rapturous applause.

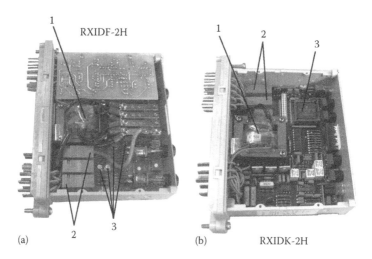

FIGURE 1.3
Two current relays with independent time delay, both with equal technical parameters, dimensions, and characteristics made in the identical standard cases COMBIFLEX® manufactured by the same company (ABB). (a) The static solid-state relay RXIDF-2H; (b) microprocessor-based relay RXIDK-2H; 1, input current transformer; 2, output electromagnetic relays; 3, transistors of the static relay and specialized microprocessor of the microprocessor-based one.

Moreover, since over the past decades billions of dollars have been invested in the ideas and technologies of the DPRs, and as it has become the source of profit for thousands of scientists and engineers all around the world, all discussions about the problems and disadvantages of the DPRs have been nipped in the bud or met with fierce opposition by the representatives of manufacturers, scientists, developers, designers, and all other participants of this business.

My past attempts to draw attention to the problems of the DPR [4,5] caused fierce accusations of incompetence, misunderstanding of the basis of the RP, and even an attempt to slow down the technological progress.

1.3 About Technological Advance

The cheapness and availability of the highly integrated microprocessors, industrial controllers and advanced electronic components, a huge and ever-expanding range of such components available on the market, extremely high performance of the equipment designed for automatic installation and soldering of the surface-mounted components of printed circuit boards, and automatic test systems for the ready-made printed boards—all these remove the previous restrictions to the complexity of the electronic systems and their field of application. This is the reason why the microprocessors can now be found everywhere, even in toilet seats, where they measure the temperature of the corresponding part of the body and control the built-in shower water heater to equalize its temperature with the temperature of the said part of the body [6].

Such a universal expanding usage of microprocessor-based electronic components in all fields of the technology together with their persistent sophistication defines the tendency of the technology development. This tendency is what we call an "advance" in the development of engineering and technology. Of course, there are certain fields of engineering and technology where computing and microprocessors are "must haves" and microprocessor technology has really enabled making a technological leap. However, implementation of the microprocessor-based devices is not always required due to the technical requirements to the products, and the number of such cases grows like an avalanche. Nevertheless, if you look at this tendency not as a bystander but as the "insider," in charge in maintenance and repair of complex industrial electrical devices, such as RP, high-power battery chargers, inverters and converters, and uninterruptible power supplies, you have to ask yourself if the tendency can really help the technical advance. You have to ask, "Why?"

Just because the current boom resulted from the sharp sophistication of the devices and ever-widening usage of microprocessors in all fields of technology and is generated by the intention of the manufacturers to outperform competitors, as well as their rush for innovations and increased profits by any and all means rather than by the real needs.

The aim to create something new or to reduce production costs could only be welcomed if this trend of substituting analog systems on discrete electronic components, which have proved their reliability for tens of years, to microprocessor based would not lead to the significant sophistication of the equipment, making it irreparable, unreliable, and expensive in maintenance in addition to the need for the sharp increase of qualification of the personnel. When you order the equipment, all these problems stay in the background, and you face them only during the maintenance. This is the price that the consumers must pay for the so-called "advance," that is, for the thoughtless and irresponsible complication of the devices, which is often proved only with the "technological fashion" and pursuit of profit. See more details on this issue in Ref. [6].

1.4 Smart Grid: One More Dangerous Vector of the "Technological Advantages" in Power Industry

You could hardly find a media that did not resound with smart grid praises. The so-called "intelligent network" or smart grid is the technological "top fashion" promising us unprecedented wealth. Today, everyone talks about his or her contribution to the development of this new fashioned trend.

1.4.1 Smart Grid Russian Style

The following is a part of the transcript of the meeting between Vladimir Putin, chairman of the Government of the Russian Federation, and Oleg Budargin, chairman of the Management Board of the Federal Grid Company Unified Energy System (JSC FGC UES), published on a website of the government of Russia (http://premier.gov.ru/events/news/9429/):

Vladimir Putin: I would like to draw your attention to the creation of the so-called smart grids.
Oleg Budargin: The work has already been started. We will not just see everything and control everything. It will diminish losses, boost energy efficiency and make power supply steady. We will get rid of risks.

While he did not mention the required amount of investments or expected economic impact across the whole power network, he reported that the required money had already been included in the 2010–2012 investment program of the company (total investment amounts to 519 billion rubles).

What is "intelligent network" or "smart grid"? Why are they going to invest so much in this technology, as it seems, while the economic impact is unknown?

An attempt to find the clear definition for smart grid, which involves such a large capital investment, surprisingly showed that no one in Russia really knows what is this all about and where this significant sum will go. Am I kidding? No! The following quotes from different specialists prove this:

> In Russia, Smart Grid technology has several alternative names – one is difficult to understand in Russian "Smart Grid," others are more descriptive – "Intelligent Power Network," "Intelligent Electric Power System" and "Active Adaptive Power Network." Presently, there are numerous definitions of Smart Grid while each involved party (such as Power Company, power consumer, power facilities Automation Company, system integrator, etc.) recognizes different functionality and tasks for Smart Grid. [7]

> Different sources define Smart Grid differently. In Russia it is known as "Intelligent Power Network," "Intelligent Electric Power System" or "Active Adaptive Power Network". [8]

> For a start, let's give the clear definition for the term of "Smart Grid." I repeatedly realize that there are serious divergences and even adverse opinions on the understanding of Intelligence in connection to power line or electrical network both between electricians and public. [9]

Let's look at the background. The first time this term appeared was in 1998 in an article of one of the Western specialists [10]. It also appeared in the title of an article by Massud Amin and Bruce Vollenberg: "Toward a smart grid" [11]. In the beginning in countries in the West, this term was used only to promote certain brands of special controllers designed for managing operating modes and synchronizing unstable voltage and frequencies of standalone wind power generators with mains. Later, this term, also only as a gimmick, was used for microprocessor-based electric meters capable of collecting, handling, and evaluating data with follow-on transmittal through communication lines or on the web. Such synchronizing controllers for wind power generators and microprocessor-based electric meters of different brands were available on the market before the appearance of the "smart grid" term. This name appeared much later as an advertising gimmick aimed at the canvassing of customers and was used only in promotional articles for the special controllers mentioned earlier.

Recently, however, it has been applied to data collection systems and the handling and monitoring of power equipment [12]:

> As a whole, the Intelligent Network (Smart Grid - "smart" or active-adaptive network) is a distribution network combining integrated tools for control and monitoring, IT and communications providing significantly higher network efficiency while enhancing the quality of energy supplied to the community by supply and retail companies and public utilities. The new distribution network will be based on following solutions:
>
> - SCADA system providing comprehensive network control
>
> - Data channel (including cable bus based on second generation PLC)
>
> - Family of teleautomatic and telecontrol digital units managing and controlling 6-20 V devices, which are installed inside medium-voltage cells during manufacture.

This is the way that the smart grid term is used in the West [13]. However, Russia, as always, goes its own way, thus enormously expanding the meaning of a well-established term.

Western smart grid is a digital technology with two-way communications encompassing all parties involved in production, distribution, storage, and consuming of energy.

Russian smart grid is an all-inclusive upgrading and innovating development of all electric power industry units based on advanced technologies and countrywide balanced design concepts [14].

Actually, in Russia, this Western term covers the whole electric power industry. So there is a logical question—why do Russian power industry bureaucrats need to extend the interpretation of this Western term so much that it loses its original significance?

The answer is cited here:

> In my opinion, our fundamental issue is that in our country the struggle of people always precedes the struggle of ideas. For state power agencies the development of Intelligent Networks means huge public finance and each department tries to "cash in on it." Also, there are numerous ideas that alike but some are "parallel-perpendicular" initiatives, which are not possible to completely align and balance together. [15]

As it happens, in Russia the smart grid concept is nothing but a battle of different industrial and power entities for the state financing.

This was implicitly proved by Sergey Shmatko, Minister of Energy of Russia, when he declared in his complimentary speech at a roundtable discussion "Smart Grid – Smart Energetics – Smart Economics" that transition to "smart energetics" will allow significantly transforming the present energy layout and push the development of power industry and implementation of innovations and new equipment at plants and by design institutes assigning an applicative meaning to developments of Russian scientists [16].

Other central figures of Russian Energetics echo this idea emphasizing that the smart grid is nothing less than the whole electric power industry of the Russian Federation:

> Smart Grid is a summation of power lines of all voltage classes, active electro-magnetic transformation devices, switching units, protection and automation equipment, IT and adaptive control systems. [17]
>
> Building of Smart Grid should be deemed as strategic lines of distribution network development and can be divided into four major fields of development:
> 1) Power equipment, technologies of energy transmission and distribution
> 2) Process management
> 3) Special communication and data devices
> 4) Automated energy accounting and control systems [7]

Certainly, there is nothing wrong with subsidizing the power industry by the government as this field really needs some innovations and the upgrading of worn-out equipment. But why the smart grid? The truth is that this popular term is a "golden key" that is able to open the door to the State Treasury. So it is used by all who are seeking a piece of the state pie, and each one does it in its own way. For example, the CEO of OJSC Electrozavod thinks [18] that smart grid is nothing but production manufactured by his plant including even furnace transformers:

- HV supply transformers and autotransformers for power stations and 35–750 kV power lines
- Shunting, current-limiting, grounding reactors and filter chokes along with other reactors of 0.5–1150 kV classes
- A wide range of 3–750 kV voltage and current transformers with enhanced precision of measurement (up to 0.2; 0.2 S)
- 10 kV dry- and oil-type transformers with a capacity of 25–4000 kVA
- Furnace transformers for electric arc furnaces, induction furnaces, electroslag remelting and ore-smelting furnaces, secondary metallurgy plants, etc.

Another author [19] considers smart grid as superconducting cables and reactive power compensators:

> Smart Grid assumes using reactive power and voltage adaptors, power concentrators, superconducting cable lines and short-circuit current limiting devices. This year the up-to-date reactive power adaptor STATCOM will be put into service at Vyborgskaya power station (400 kV) to improve

the steadiness of electric energy exported to Finland. An asynchronic reactive power compensator will be activated at the 500 kV Beskudnikovo substation for maintaining optimum voltage and increasing network transfer capacity. This will result in a steadier power supply for consumer from North and North-East regions of Moscow city.

Those are just a few examples of the numerous attempts of different entities to wheedle money out of the government to finance their own projects under the guise of the smart grid. The issue is whether the investment program will be capable of covering everything—from furnace transformers and superconducting cables to reactive power compensators—or, as usual, this whopping sum will be ladled out and the broadcast campaign will end until the next hot overseas byword makes an appearance.

1.4.2 Smart Grid: Western Style

While the "smart grid" term is twisted in Russia, in the West, this term is misused as well. In recent years, this term was significantly transformed and lost its original meaning. Today, it covers almost all of the power industry—from power production systems and grid structure and configuration to metering and information-measuring systems, automated control systems, communication between power facilities, and RP. As we can see, today, this term has no clear definition and is used differently by different authors, so it is not possible to get a clear understanding on the root of the smart grid. Articles of some authors where the term smart grid is used in the header are focused on construction principles and configuration features of power mains, while others on establishing communication channels and data transmitting principles or on environmental issues and alternative power sources. As is readily apparent, the application of this term has been made absolutely senseless.

Usage of this term in different national programs on reconstruction and modernization of power industry is also senseless as such major programs last for decades within which equipment and technologies are gradually changed and require enormous investments made possible only in portions for individual projects. So the reality is that today we are only able to consider individual components such as the formidable concept of energetic development as the smart grid rather than using the old, well-established and common terms with clear definitions.

1.4.2.1 Power Generation Systems

Climate change issues and predicted lack of organic fuel promote the development of alternative sources of electric energy, such as wind generators, solar photovoltaic systems, biofuel generators, tidal wave power generators, and geothermal power generators. Pumped storage developments enhancing efficiency of produced energy use are another trend. It is expected that in the future the number of such sources will be connected to different points of a common network. This means that in the future, power production facilities will be distributed rather than concentrated, as they are today. The important feature of such sources is relatively low capacity and instable power parameters. It is clear that in order to stabilize the parameters of such sources and provide their automatic synchronization with the grid, an intelligent control device is required.

The development of brand new power generator systems and improvement of technical and economic efficiency of existing ones, designing appropriate automatic control devices and communications systems providing information exchange between such sources and other power system units, are some of the smart grid concept directions.

1.4.2.2 Electrical Grids

Today, electrical grids are hierarchic (generator, main lines, distribution networks, municipal networks, etc.) [20] (Figure 1.4).

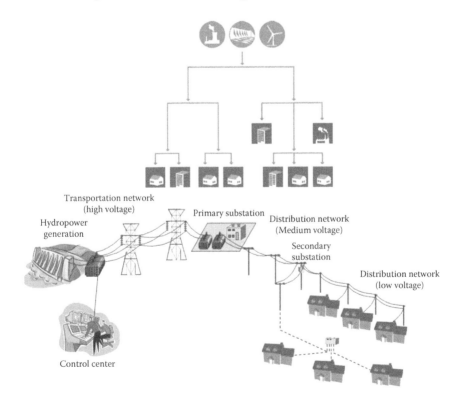

FIGURE 1.4
Structure of a traditional hierarchic electrical grid.

Most modern electrical grids consist of radial lines with one-way flow of energy. Only some of them are completed. Future smart grid networks will not be hierarchic, while large consumers will be "mixed" with a large number of relatively low-power energy sources, individual high-power plants, voltage regulators, reactive power compensators, etc. It will be a very complex, unstructured, and extended network.

Within this network, power flows will not be strictly deterministic. It is obvious that such a complex unstructured grid (which can, in a sense, be compared to the Internet [20]) should have a powerful control system to align the operation of all network components with each other. It demands all components of the network to "dialogue" with each other and with the control center through specific communication networks that are supposed to be wireless. The development of powerful fully managed network components equipped with self-diagnostic systems and monitoring capabilities, as well as with reliable data transmission and input channels, is one of the directions of the smart grid concept development.

1.4.2.3 Monitoring and Self-Diagnostic Systems for Electrical Equipment

Both high sophistication of powerful grid components and progress in the computer-controlled systems intensify further development of the health monitor systems for the electrical equipment providing early failure prevention on the important components of a network. The law of the aging of electric isolation, knowledge of the trends in the composition of supply transformer oil, and known features and properties of partial discharges in solid, liquid, and gaseous isolation and in vacuum enable creating special indicators and reliable diagnostic procedures for the constant monitoring of the health of important network components constituting another part of the smart grid concept.

1.4.2.4 Communications and Data Transfer across the Electric Power Facilities

Today, communication and data transfer across the power facilities are realized through different circuits. This includes low-voltage (LV) communication networks (low-frequency control cables, coaxial high-frequency cables), optical cables, HV power lines, and protected directional radio channels. In recent years, such network technologies as the Ethernet/Internet have become increasingly applicable. This primarily resulted from their cheapness, prevalence and ubiquity, and well-developed technology and communications protocols, as well as from expected jumps in the sizes of data files exchanged across multiple power network components scattered over a large territory. Even now the market offers various electronic sensors, transducers, and sensing devices equipped with cheap built-in Ethernet/Intranet modems. Optical communications used in today's RP is considered to be too expensive for the extended and universal application within the future smart grid concept [21]. However, this process is accompanied with a great deal of fluff as different companies occupying particular sectors of communication and data market try to validate usage of their principles and data systems within the smart grid concept. For example, along with allegations that the future belongs exclusively to the standard network applications, such as the Ethernet/Intranet, there are statements that the only valid communication is a broadband HV power line solution [22]. Also, there are quite serious discussions on the application of habitual and common wireless communications such as cellular networks, WiMAX, and Wi-Fi, within the smart grid [23] (Figure 1.5).

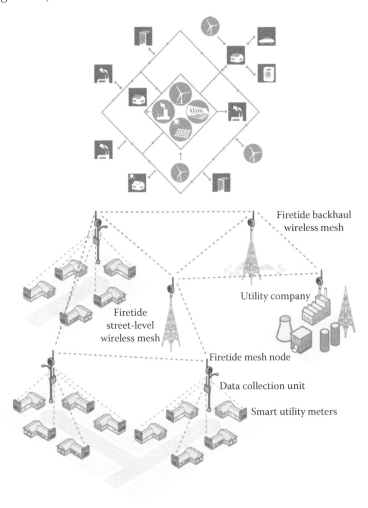

FIGURE 1.5
Smart grid structure with wireless information control network.

1.4.2.5 Electric Power Metering System

Microprocessor-based power meters entered the market many years ago without any connection to the concept of smart grid (Figure 1.6). On the contrary, the purely promotional term "smart grid" initially appeared only to promote such meters that grew up in a kind of global vision of the future of power industry. Multirate microprocessor-based meters capable of performing calculations, communicating with other similar meters, and accumulating and transmitting data over the network have been used in energetics for many years. In recent years, simplified versions of such meters have been applied in everyday life. Now the level of the art achieved in this area fully complies with the concept of smart grid.

1.4.2.6 Smart Grid Operating Principle

In accordance with Ref. [24], the reliable operation of such a complex system as a smart grid can be achieved by minimizing the number of individual multifunction data process modules (this means further concentration of functions in single modules). Data sent from multiple components of the smart grid must be transferred to powerful servers through networks, processed by computer centers, and sent through the network to actuators. According to Ref. [24], all the basic functionality of the smart grid should be provided at the software level.

1.4.2.7 Technical and Economical Aspects

The smart grid is a concept associated with the global reconstruction of the whole power supply system. It is obvious that implementation of such a global program demands enormous investments. Thus, it is logical to ask how, in fact, it will benefit us. What economic returns can we expect from these investments? Unfortunately, none of the numerous publications describing the advantages of smart grid, as we found out, gives any business case for the realization of the smart grid concept. Doesn't

FIGURE 1.6
"Smart" electric power meters.

the existing structure of electrical networks provide a steady electrical supply for consumers? Aren't the microprocessor-based electric power meters more widely applicable beyond this concept? Does the development of modern microprocessor-based automated diagnostic systems suffer from the lack of the smart grid? Aren't modern DPRs capable of meeting all current challenges of relays? Certainly, the radical change of network configuration and appearance of numerous power sources in a network can change functions and algorithms of RP dramatically. However, how can nationwide electrical network structures, which have developed for decades, change so fundamentally in rather practical than theoretical ways? And what for? As for a large number of small power generating sources (wind generators, solar batteries) that the smart grid apologists expect to be included in the future general electrical network, such developments are doubtful. As we saw in many European countries (Italy, Netherlands, Germany, Spain, etc.), wind generators or solar batteries were not used (as power network entities) as single network power units (except for the individual devices powering separate facilities). Common practice is to combine them in the large power units occupying huge areas (see Figure 1.7) and connected to grids. For example, the capacity of the Thanet wind generator located at southeast coast of Kent county in Great Britain and consisting of 100 wind turbines (the planned number is 340) is 300 MW.

Available microprocessor-based automatic operation systems successfully manage such large power installations and synchronize them with the networks without any "smart grid." Besides, as we know, the wind power industry is not all that profitable. According to the experts of the British Energy Research Centre, the energy produced at the coastal wind power stations is approximately 90% more expensive than the energy produced from traditional fuel sources and 50% more expensive than the energy produced from nuclear fuel.

On the other hand, if we consider the smart grid concept as a fundamental reconstruction of electrical grids resulting in the significant sophistication of their structures and operations, we should be aware of the predictability of such modes and the ability to determine who should calculate RP set points and how for such complicated networks. Another point is to what extent the set points will be able to reflect the actual grid emergency modes. We expect that due to both the complexity of the network and a high number of cross-coupled active components, it will be a real challenge to find the reason a failure occurred, even using the self-diagnostic devices. It will require modeling of network operation modes along with considerable research.

Compared to the existing grids, we assume that such networks will be much more difficult to operate and require far more skilled staff. The "one-and-all" computing process covered by the smart grid concept is already in full play in both industry and energetic. The absolutely outrageous willingness to integrate all kinds of power equipment with the computer network and all

(a)

(b)

FIGURE 1.7
(a) Modern solar (photoelectric) and (b) wind generator stations.

round and to move from the old reliable analog electronics to digital microprocessor-based units very often results in catastrophic consequences [6].

1.4.2.8 Smart Grid: Panacea or Road to Hell?

As is known, the smart grid concept assumes the installation of microprocessors in absolutely all elements of the power production, distribution, and metering systems as well as arrangement of data channels between them based on the computer networks, which are generally wireless (Wi-Fi). In fact, if all elements of the smart grid are to be controlled by commands through the networks with TCP/IP protocols, there is an enormous potential danger of external intervention to the power system operation. Many experts emphasize this hazard [25–46] devoting international conferences [47] to it. Only apologists of the smart grid "do not notice" these problems. What do we hear from apologists of the smart grid? Nothing, we hear only the usual reservations about the necessity to isolate the internal network of the smart grid from the external web and about access passwords and other trivial safety measures. We all understand that all these measures can limit access for normal people, but not for experienced hackers cracking even the very well protected networks of the Ministries of Defense and banks.

The apologists of the smart grid see the future power system as a fancy modern online game with thousands of participants—the components of the power networks. If we also consider the millions of the household power meters united in a common computer network (i.e., millions of potential points of network access for hackers), the grandeur and the danger of this idea due to the sharp increase in vulnerability of the power systems to hacker attacks, computer viruses, and remote intentional destructive electromagnetic impacts becomes even more obvious.

Today, the low-power, high-altitude electromagnetic pulse of a nuclear explosion set off in near space over the territory of a country is regarded as a real type of the so-called nonlethal weapons capable of disabling all the microelectronic devices over the whole country while not injuring people. Alas, all these dangers, or "horror stories" as they are disparagingly referred to by some proponents of the "technological advance" in its present form, are hardly considered by scientists and engineers who receive their payrolls from the funds allocated to the development of the smart grid. They often say our task is to advance the technologies, while protection of the national power systems is the "headache" of the army and intelligence agencies. Defectiveness of such an ideology is obvious and does not even require an explanation.

So what really is the smart grid? In our opinion, the smart grid is a global publicity campaign juicy for hundreds of manufacturers, research centers, and universities. Its purpose is to promote numerous products, technologies, and researches compulsorily "attached" to the popular term of "smart grid" and "wheedle" enormous investments out of the state budgets [48] to develop and manufacture products and systems under the devastatingly extended term of "smart grid." As for apologists of the smart grid, it seems like their financial interests prevail over the sensible doubts and fears concerning the dangers resulted from the full implementation of the concept. Neglecting these hazards can lead to global international catastrophes.

Moreover, the normal growth of technology will not be slowed down if the concept isn't implemented. It will only be sounder, more rational, and more careful.

1.5 Dangerous Tendencies in the Development of the Relay Protection

There was an interesting story in an old science fiction novel. It all started from such innocent thing as an odd night call to all the phones on planet earth. The call announced the birth of global intelligence to all people of the earth. It turned out that at some stage of development the proliferation of computers escalated into a new quality: the millions of computers, integrated into a single network that controlled everything and everyone on the planet earth, suddenly became conscious of themselves as an entity capable of reproducing itself through the automated factories and robots connected to the same network, as well as of defending itself with computerized weapon systems designed to destroy the human race. From the perspective of the global intelligence, the humanity was nothing but a useless vestige, gobbling up the planet's resources. There are no prizes for guessing about the further development of the action.

Network-connected computers already control almost all types of modern industrial production systems, water supply and electricity systems, and telecommunications systems and networks. New terms such as smart grid and RP with artificial intelligence have emerged in technical literature rather than in science fiction. Today, technical literature rather than science fiction refers to the creation of so-called smart house, where even the refrigerator will analyze the stored products, and, based on the analysis of consumption, will make an order and send it over the network to the nearest supermarket. Today, you can find microprocessors everywhere, even the water closet lid [6].

Humanity is moving by leaps and bounds to the creation of all-powerful global intelligence prophesized in the old science fiction novel.

But let's get back to reality. And the reality is that major failures in the energy systems that have occurred in America and Europe (United States, 1965, 1977, 2003; France, 1978; Canada, 1982, 2003; Italy, 2003; London, 2003; Sweden, 1983, 2003) were caused by incorrect or, rather, unpredictable actions of RP during complex emergency modes due to disabling the wrong sectors of the network in every particular situation. Had the action of RP under these specific circumstances been different, the system failures might have been avoided. I am referring to the power systems (United States and Western Europe) that have already been equipped with computers and microprocessor-based protection. For comparison, I should mention that one of the world's largest power systems with negligible percentage of computerized RP devices and the old worn-out equipment has never suffered from such failures. I am referring to the power system of Russia. The answer to the question about the reasons for this can be found in the book by E. M. Schneerson, Doctor of Technical Science and foremost authority in the field of modern RP [49]:

> Improvement of the technical level of relay protection devices (RPD) alone does not necessarily lead to the equivalent improvement of efficiency as related to the response to emerging damages. For example, outdated electromechanical and to an extent electronic static RPD, if protection functions and settings are chosen correctly, certainly provides better protection for the network than the microprocessor RPD without rational definitions of the specified parameters.

In previous publications, we repeatedly drew attention to the dangers of some tendencies in the development of RP that is strongly promoted by the developers and manufacturers of the DPRs:

1. Continuous sophistication of the DPRs and increasing the number of protective functions in a single terminal [50,51].
2. Overloading the DPRs with functions unusual to the RP, such as the monitoring of electrical equipment [52,53].
3. Use of a nondeterministic logic in DPRs, as well as so-called "advance actions" that lead to the risk of the loss of control over the RP actions [53,54].
4. Wide use of free programmable logic [54] in DPRs, resulting in the significant increase in the percentage of staff mistakes and faults of protection.
5. Complication of the serviceability checks and maintenance of the relays, while integrating numerous DPRs of different types and brands with different designs and software in the same power network. The lack of the common standards for the DPD design and software increases the intellectual load on staff and leads to significant economic losses [55]. This situation is exacerbated with every passing year.
6. The dramatic weakening of the electromagnetic immunity of the RP and the whole power system in proportion to the usage of the DPRs [56–60].
7. Increased vulnerability of the power systems to hacker attacks resulting from the expansion of the microprocessor-based devices and the usage of the cheap Ethernet and Wi-Fi lines instead of the relatively protected optoelectronic cables in the RP systems [61].

In fact, the behavior of electromechanical and electrostatic RP devices under emergency situations was rigidly determined by their operating principle and settings. Current trends in the development of DPR are associated with the increase of their "independence" (i.e., in fact, unpredictability) in making decisions. It relates to the RP self-learning capability peculiar to adaptive neural networks, as well as the use of technologies of artificial intelligence with fuzzy logic.

Another clear trend in the development of the modern DPR is the excessive complexity by including extraneous functions not typical for RP. Here, for example, is the list of functions performed by the so-called "intelligent controller" (Intelligent Protection and Automation Controller [iPAC] by Dynatrol Systems Inc.):

IEEE protective functions:

- Sync check (25)
- Under voltage (27)
- Directional power (32)
- Phase balance (46)
- Instantaneous overcurrent (50)
- Inverse time overcurrent (51)
- Overvoltage (59)
- Voltage balance (60)
- Directional (67)
- Reclosing (79)
- Under and over frequency (81)
- Lockout (86)
- Differential for transformer protection (87)

Measurement functions:

- Voltage and current rms values
- Neutral current rms values
- Power factor measurement (power factor correction: capacitor bank switching)
- Total power measurement
- Real and reactive power measurement
- Power quality measurement (fast Fourier transform (FFT) for harmonic measurement)
- Frequency measurement
- Total harmonic distortion measurement

Advanced features for protections:

- Coil monitoring for relay failure detection
- Cold load pickup logic to prevent protective devices from operating when cold load is put on the circuit

- Voltage constraint with current pickup lowered to increase sensitivity when voltage is also collapsing during the fault
- Breaker control blocking for coordination with upstream and downstream protective devices via DI or peer-to-peer communication
- Directional on overcurrent devices

Add to this the monitoring of external current and voltage circuits, the registration of events, functions of emergency digital oscilloscope, and other routine functions of the DPR.

Then there is the danger of excessive concentration of protective functions in a single terminal (e.g., a microprocessor relay–type SYMAP by Stucke Elektronik GmbH incorporates 39 separate protection functions); additional RP functions, extrinsic to the protection itself, lead not only to the physical complication of the device, consequently reducing its reliability, but also to the complication of its software and user interface. This in turn leads to a sharp increase in the number of software errors (the so-called "human factor"). Due to such large number of functions, using the same internal resources of DPR and possible conflicts of the embedded logic functions during complex emergency mode accompanied by a transition of one type of damage to another, it is not always possible to predict the behavior of the protection. Damage to one function that is common to all the internal element DPR functions (power supply, watchdog, memory, microprocessor, its servicing subassemblies, etc.) will result in the instant failure of all protective functions at once.

Despite the obvious problems existing today due to the excessive concentration of protective functions in a single terminal, some leading experts in their philosophizing about the future of RP not only advocate additional "adding on" of extraneous functions to RP but even go further.

They put forward the fantastic idea of so-called "multi-dimensional relay protection" [62] and relay protection with proactive functions [63], acting on the basis of its own experience, its own analysis of the current status of the protected object, and prediction of its future state. In essence, this is about the RP capability taking completely unpredictable actions, as an independent intelligence making its own (previously not determined) solutions and changing power systems operating modes (through switches) at its own discretion before the emergency mode occurs [64].

It must be emphasized that there is nothing wrong with the development of computer-based diagnostic and prediction methods for electrical equipment, and it could only be welcomed but for the attempts to "intercross" it with the protection relay.

Apart from the risk of losing control over the RP actions, current trends of its development dramatically increase the risk of hacker attacks on the grid as a computerized RP system is a good target for changing the state and affecting the modes of power systems. Despite the serious concern of specialists about this problem [64,65], the trend toward greater susceptibility of power systems to hacker attacks only grows.

Another serious threat to the stability of power systems, which is based on current trends in its development, is the development of technologies of artificially produced destructive impacts on electronic and computer equipment [66–72]. The development of these technologies throughout the world contributes to increasing the spread of microprocessor technology and the memory elements with high sensitivity to external electromagnetic emission, on the one hand, and the tendency of constantly increasing the density of microelectronic components as a result of the reduction of the thickness of the operating and the insulating layers in the crystals, on the other hand. These two tendencies, directed toward each other, form a very dangerous vector of development of modern technologies. Moreover, today, you do not need any special knowledge or equipment in order to create a device capable of destroying all the electronic devices of your neighbor.

This sophistication, of both hardware and software, has not sunk in. As shown in [51,73–76], even now the transition to the DPRs is accompanied by a significant decline in the reliability of the RP. However, the advocates of microprocessor-based RP believe that we should not be satisfied with what has already been achieved and must further sophisticate the DPRs by increasing the number of functions performed by a single terminal, by using freely programmable logic and nondeterministic logic based on the theory of neural networks and algorithms of preemptive action, by overloading the DPRs with the functions of information-measuring systems and monitoring systems of power equipment, and by using the wireless communication channels (Wi-Fi) between relays.

The new-fangled ideas and developments in the field of DPRs are no longer limited only to the functions of RP. It is supposed that in the near future not only RP but the whole power systems should correspond to the smart grid concept, which implies that all the power equipment of power system elements must be based on microprocessors to manage the exchange of synchronization signals and control commands between such elements via Wi-Fi.

What alluring prospects and inviting horizons! What enormous amounts of money are to be allocated from the government budgets to the new programs in the power systems! So many research and production teams can subsist on these budgets, periodically suggesting more and more improbable but very "beautiful" ideas and putting on the market more and more sophisticated but less reliable products.

It's a huge business, and nobody wants to be banned from this sweet "cake." The participants in this business are not concerned about the latest affects of their activities and seek only to quickly "push" their new-fashion ideas to the market.

Business is business and its fundamental laws are the same in all countries and areas, including such sensitive ones as RP and power management

by Bernd Michael Buchholz, NTB Technoservice, Germany and
Christoph Brunner, it4power, Zug, Switzerland

industry reports

and prosperity of the industry was clearly considered by Madame Merce Griera I Fisa from the European Commission. The SmartGrids are a prerequisite to reach the European 20-20-20 targets in 2020 (20% *improvement of energy efficiency, 20 % share of renewable energy sources to cover the demand of primary energy, 20 % reduction of carbon emissions*). Furthermore, the advanced products and system solutions partly resulting from funded projects will ensure success of the European industries

presented by the 12 participating project teams – beginning with the building automation "SmartHome" and the involvement of household consumers into the electricity market, the automation of distribution networks up to the erection of prospective markets for energy and reserve power. Engaged discussions followed each of the contributions.

The analysis of the consumer behavior in the environment of dynamic tariffs presented a potential of 14% energy saving and load

It is mandatory that the new solutions from the project are urgently applied in practice now.

One session considered the barriers for SmartGrid solutions by the current regulation and legal situation in Germany. For many years the German Power Engineering

FIGURE 1.8
The motto to one of the articles in the worldwide popular specialized magazine *Protection, Automation and Control Magazine—PAC World*, September 2011 (in the border).

and control. Do you need a proof? Read the motto to the report on the symposium "Distribution systems of the future: Novel ICT solutions as the backbone for smart distribution" published in the *PAC World Magazine* (Figure 1.8).

The keywords in this motto are "urgently" and "mandatory," that is, without the careful analysis of the remote effects of these innovations and without "unnecessary" criticism. Thus, it has worked until today it is worldwide.

There was a period of sharp criticism of my publications and total negation of the nonamenities of the aforementioned relay development tendencies. However, in recent years, many experts have started to understand the problems I have revealed (and, generally, without any reference to my earlier articles in major technical magazines of Russia, Europe, and the United States). For example, at the *Second International Conference: "Actual Trends in Developments of Power System Protection and Automation"* (Moscow, Russia, September 7–10, 2009), B. Morris, R. Moxley, and C. Kusch (Schweitzer Engineering Laboratories, United States) submitted a report "Then versus now: A comparison of total scheme complexity," where they questioned the need for further sophistication of the protection and appealed to the comparative assessment of the reliability of the simple electromechanical RP and multifunction microprocessor-based protection systems. They stated that they have identified a trend of downward reliability of the RP systems built on sophisticated microprocessor-based devices.

The unreliability of DPRs was also raised by V. I. Pulyaev (FSK UES, Russia) at the *Third International Conference: "Actual Trends in Developments of Power System Protection and Automation"* (St. Petersburg, May 30–June 3, 2011). He particularly noted the significant failure rate of relaying accounts for microprocessor units (approximately 23% of all cases), which constitute only about 10% of the total number of protection devices. This is definitely one of the most important factors determining the need for special measures to enhance the reliability of the DPRs.

Now late Alexey Shalin (PhD, a professor of Electric Power Stations Department of the Novosibirsk State Technical University, leading specialist of LLC "PNP BOLID," Novosibirsk), in his article responding to one of my publications (see Ref. [77]), commented that the percentage of malfunctions of the modern relay panels and cabinets often were significantly higher than of the old defenses based on electromechanical relays. He also presented statistical data confirming that the transition from the defenses based on electromechanical relays to microprocessor terminals was accompanied with a significant reduction in the efficiency and reliability.

The unreliability of current DPRs was the focus of the articles by A. N. Vladimirov (Central Dispatch Administration of UES of Russia), S. Swain, D. B. Ghosh (Integrated Electrical Maintenance), and others [78].

J. Stokoe and J. Gray in their report "Development of a strategy for the integration of protection & control equipment" submitted at the *7th International Conference: "Developments in Power Systems Protection"* (Amsterdam, the Netherlands, April 9–12, 2001) pointed out that the old electromechanical relays were strong and durable devices with a lifetime of 25 years, whereas the life of modern microprocessor-based protection is 15 years or less.

They echoed by J. Polimac and A. Rahim (PB Power, United Kingdom) who declared that the transition from electromechanical relays to microprocessor-based ones reduced the lifetime of protection from 40 years (EMR) to 15–20 years, and sometimes even to only a few years after commissioning (DPR) [78].

The head of the Computer Department of the Engineering and Technology College (University of Poona, Maharashtra, India), Ashok Kumar Tiwari, B. E., noted that the integration of numerous functions in a single microprocessor terminal significantly reduced the reliability of the RP since the failure of the terminal will result in the loss of too many features compared to the system where these functions are distributed among several terminals [78].

The necessity to limit the number of functions combined in a single DPR terminal was also mentioned by V. A. Efremov, S. V. Ivanov (IC "Bresler"), and D. V. Shabanov (Russia FGC) in the report "Actual trends in developments of power system protection and automation" also submitted at the same *Third International Conference*, mentioned earlier.

A. Fedosov and E. Pusenkov (a subsidiary of OAO "SO UES" ODE, Siberia) in their article [79] noted the lack of the strict standards on the DPR hardware and software has resulted in too great a variety of programs and algorithms built in the power system DPR, which has led to the faulty operations and increased the likelihood of faults of such devices.

The sharp vertical growth of the tasks performed by the personnel servicing the RP after the transition from EMR to DPR was referenced as the cause of severe accidents in power systems by D. Rayworth and M. A. Rahim (PB Power, United Kingdom) [78].

A. Belyaev, V. Shirokov, and A. Emelyantsev (Specialized Department "Lenorgenergogaz," St. Petersburg) in their article [80] also considered the complexity of program interface and the necessity for inputting numerous set points during the programming of DPRs.

The poor electromagnetic environment in most of the old substations designed and built for the electromechanical RP, and not for microprocessor-based devices, as well as the resulting numerous MPD faults, were noted by B. I. Kovalev and I. E. Naumkin (Siberian Energy Research Institute), A. M. Bordachev (JSC "Institute Energosetproject"), M. Matveev and M. Kuznetsov (OOO "Aesop"), P. Montignies and B. Jover (Schneider Electric, France), V. Nadein ("Arkhenergo"), V. Lopukhov (SUE "Tatenergo PEO"), A. Ermishkin (JSC "Mosenergo"), R. Borisov (NPF "ELNAP," Moscow), A. W. Sowa and J. Wiater (Electrical Department, Białystok Technical University, Poland), and others.

Many experts noted that the vulnerability of DPR to the electromagnetic interference is several orders higher than that of traditional electromechanical counterparts, and therefore to ensure the electromagnetic compatibility (EMC) of the secondary circuit, their level of electromagnetic protection has to be significantly higher. Thus, an acceptable level of DPR reliability can be reached only after providing for their EMC. Low EMC survivability of the DPRs is closely related to the deeper and more dangerous problem of remote intentional destructive electromagnetic impacts on the DPR.

So today, we should stop turning a blind eye to the tangle of problems associated with the proliferation of the DPRs, disparagingly calling them "scary stories from Gurevich," since already today dozens of experts from many countries state that there are serious problems that need to be addressed.

1.6 What to Do?

As shown [4,51,73–76], the reliability of DPRs is lower than the reliability of electromechanical relays even today. However, it doesn't mean that it is necessary to slow down the transition from EM relays to DPRs. Rather it sets a serious challenge to be resolved. Some lines of attack on the problem have been proposed by the author [2]. In a few words, they can be combined as follows:

- Do not flood DPR with the functions beyond RP such as monitoring of electrical equipment.
- Limit the number of functions in a single microprocessor terminal; optimize this number not only by relaying cost but also by its reliability.
- Refuse the fuzzy logic algorithms providing RP unpredictability.
- Simplify the program interface as far as possible based on some unified DPR software platform.

- Leading manufacturers of computer-based test equipment for DPR should develop a set of programs fully compatible with the unified DPR software platform and allowing complete automation of DPR testing to minimize human errors.
- Develop new DPR design principles based on universal interchangeable functional modules such as personal computers (PCs); create the market of universal functional DPR modules.
- Carry out special research and development efforts for ensuring required functionality of RP under malicious destructive electromagnetic impacts.

As you can see, the preceding principles are opposite to the trends of the smart grid concept. What does this mean? It means that implementation of the concept will lead to a critical decline in the reliability of RP.

In our opinion, the only purpose of RP should be the RP itself (i.e., detecting emergency modes and issuing command to the power control devices changing the operating mode of the protected object in order to eliminate the emergency mode) and nothing more. All other problems should be solved by other systems independent of the RP. On the other hand, recently more and more complicated and sophisticated systems for monitoring electrical equipment modes based on the continuous monitoring of the electrical characteristics (tangent of dielectric loss angle, partial discharges in insulation, arrester leakage current, the number and composition of gases dissolved in transformer oil, etc.) have emerged with the ability of predicting the processes that progress in time. Automatic process control systems have become more complicated as well as real-time control systems for measuring vector values of current, voltage, and power, the systems for registration and oscillographic testing of emergency operation, etc. In contrast to the RP, all these systems don't have a direct influence on the operating mode of power systems, and therefore there are no restrictions on the trends of their development.

In contrast to isolated measuring and monitoring computer systems, the protective relay is associated directly with the possibility of destructive impact on power system modes. This is the most important and fundamental difference of RP from all the other computerized devices and systems used in electric power engineering, preconditioning a need for a different approach to RP.

Therefore, further development of microprocessor RP and other microprocessor and computer systems in power engineering should take place in independent unrelated parallel courses.

In order to prevent the arbitrariness of manufacturers imposing more and more "advanced" and less reliable DPRs to the energy sector, it is necessary, in our opinion, to formulate the basic requirements for the DPR design principles (not for the technical parameters but for the principles of design) in the relevant standard. The same principles could also include earlier proposals on DPR design as of a set of individual replaceable modules universal in functions, size, and contact connections printed circuit boards (PCBs) by analogy with PCs.

In my opinion, it's time to put an end to the uncontrolled development of unproven and dangerous trends in the RP and automation systems, including the smart grid.

This requires the establishment of the National Coordinating Councils for the RP and intelligent networks, which must analyze current trends, develop national strategies, and coordinate the standardization in these areas.

The councils should include independent experts and specialists in the field of RP, microelectronics, data protection, and EMC, who do not have economic ties with the development and production of the DPRs or the elements of the smart grid. It should be noted that a purely mercantile financial interests of individuals and even entire scientific and industrial groups, interested in the funding of any new digital technologies in the power sector and, in particular, in the field of the RP, regardless of the long-term effects of such technologies and not limited to any frameworks, may result in national disasters in the near future.

New technologies in the field of RP, automation, communication, and data transfer systems should not be introduced into service until the possible negative consequences of their wide distribution are fully considered in the light of accumulated experience and until the effective measures to protect against remote intentional destructive impacts, whether intentional electromagnetic impacts or hacker attacks, are developed. The development of measures to protect sensitive electronic equipment of power systems against intentional destructive impacts should be considered as one of the main targets, and it should be funded by amounts not less than those spent for the development of new technologies, such as the smart grid. Development of any new technology, based on digital microelectronics, should be considered as complete and ready for use in power industry only after the development of measures to protect it from the intentional destructive impacts.

New standards on microprocessor-based RP, which are required according to Refs. [81], must include the requirements for the protection against intentional destructive impacts, as the current standards on EMC consider only the stability of the equipment against natural effects, rather than against intentional destructive electromagnetic impacts.

Only electromechanical relays are fully resistant to cyber attacks and intentional electromagnetic influences, requiring no live power, and thus being always ready to work can be used as such standby RP sets. Therefore, in our opinion, it is too early to dismiss the electromechanical relays. Rather, they should be improved through the new technologies and materials, and their range should be updated.

I must carefully examine the ways to improve the reliability of the DPRs by means of the modern redundancy hybrid relays [82,83].

I am sure that it is the only acceptable direction of the technical advance in such an important and basic sector of any national infrastructure as power industry.

References

1. Gurevich V. I. "Protective relay" and "relay protection": Problems with the terms. *Electric Power's News*, 4, 2012, 23–33.
2. Gurevich V. I. *Digital Protective Relays: Problems and Solutions*. Taylor & Francis Group, Boca Raton, FL, 2011, 404pp.
3. Usova S. V. A. "Bright tomorrow" will predict by dollar exchange rates. *Russian Power and Industry*, 8 (24), August 2002.
4. Gurevich V. Microprocessor protection relays: New prospects or new problems? *Electrical Engineering News*, 6 (36), 2005, 57–60.
5. Gurevich V. Microprocessor protection relays: Alternating view. *electro.info*, 4 (30), 2006, 40–46.
6. Gurevich V. I. Price for "the progress." *Components and Technologies*, 8, 2009.
7. Volobuyev V. V. What is the smart grid? What are the prospects for the development of smart grid technology in Russia? http://www.rsci.ru/sti/3755/208683.php.
8. Yegorov V., Kuzhekov S. Intelligent technologies in distribution system. *EnergoRynok*, 6, 2010.
9. Osika L. Smart grid: Expert comments. *EnergoRynok*, 6, 2010.
10. Janssen M. C. The smart grid drivers. *PAC*, June 2010, 77.
11. Amin S. M., Wollenberg B. F. Toward a smart grid. *IEEE P&E Magazine*, September/October, 2005.
12. "Smart Grid" is to be built in the center of St. Petersburg. *Reporter Magazine*, May 6, 2010. http://newsdesk.pcmag.ru/node/24272.
13. Gellings C. W. *The Smart Grid: Enabling Energy Efficiency and Demand Response*. CRC Press, Boca Raton, FL, 2010.
14. Budargin O. Smart Grid as the basis for the innovating economics. Roundtable "Smart Grid – Smart Energetics – Smart Economics," *International Economic Forum*, St. Petersburg, FL, June 17, 2010. http://www.fsk-ees.ru/evolution_technology_seminars_forum_conf.html.
15. Osorin M. Smart grid: Expert comments. *EnergoRynok*, 6, 2010.
16. Complimentary speech of Minister of Energy of Russian Federation. Roundtable "Smart Grid – Smart Energetics – Smart Economics," *International Economic Forum*, St. Petersburg, FL, June 17, 2010. http://www.fsk-ees.ru/evolution_technology_seminars_forum_conf.html.
17. JSC FGC UES discussed concept of Smart Grid building with members of the Academy of Sciences of Russia. NT-Inforf. http://www.rsci.ru/sti/news/208892.php.
18. Makarevich L. V. HV electrical equipment for development of "Smart" Unified Energy System of Russia – Roundtable "Smart Grid – Smart Energetics – Smart Economics," *International Economic Forum*, St. Petersburg, FL, June 17, 2010. http://www.fsk-ees.ru/evolution_technology_seminars_forum_conf.html.

19. Nikolayev V. Intelligence is the thing of the future. Innovating systems come to the power lines. *Independent Newspaper*, March 23, 2010.
20. *The Smart Grid Reliability Bulletin*. ABB White Paper, North American Corporate Headquarters, 2009, 14pp.
21. Shono T., Fukushima K., Kase T., Sugiura H., Katayama S., Tanaka T., Beaumont P., Baber G. P., Serizawa Y., Fujikawa F. Next generation protection system over Ethernet. *The 10th IET International Conference (DPSP 2010) on Developments in Power System Protection*, Manchester, U.K., March 29–April 1, 2010.
22. Renz B. Broadband over power lines (BPL) could accelerate the transmission smart grid. DOE/NETL-2010/1418, National Energy Technology Laboratory, U.S. Department of Energy, Washington, DC, 2010.
23. Why the smart grid must be based on IP standards. http://blog.ds2.es/ds2blog/2009/05/why-smart-grid-must-use-ip-standards.html.
24. Baldinger F., Jansen T., Riet M., Volberda F. Nobody knows the future of smart grid, therefore separate the essential in the secondary system. *The 10th IET International Conference (DPSP 2010) on Developments in Power System Protection*, Manchester, U.K., March 29–April 1, 2010.
25. Kawano F., Baber G. P., Beaumint P. G., Fakushima K., Miyoshi T., Shono T., Ookubo M., Tanaka T., Abe K., Umeda S. Intelligent protection relay system for smart grid. *The 10th IET International Conference (DPSP 2010) on Developments in Power System Protection*, Manchester, U.K., March 29–April 1, 2010.
26. Robertson J. Security experts offer caution on smart grid. Associated Press, New York, July 31, 2009.
27. Krebs B. "Smart Grid" raises security concerns. *The Washington Post*, July 28, 2009.
28. Slocum Z. Report: Smart-grid hackers could cause blackouts. Cnet.News, March 21, 2009. http://news.cnet.com/8301-1009_3-10201651-83.html.
29. Baldor L. C. New threat: Hackers look to take over power plants. Associated Press, New York, April 8, 2010. http://www.boston.com/business/articles/2010/08/03/computer_hackers_look_to_take_over_power_plants/.
30. Nakashima E. Defense official discloses cyberattack. *The Washington Post*, August 25, 2010.
31. Gorman S. U.S. plans cyber shield for utilities, companies. *The Wall Street Journal*, July 8, 2010.
32. Lemos R. Hacking the Smart Grid. *Technology Review*, April 5, 2010.
33. Mills E. Experts warn of catastrophe from cyberattacks. InSecurity Complex, February 23, 2010. http://www.cnet.com/news/experts-warn-of-catastrophe-from-cyberattacks/.
34. Aitoro J. R. Energy set to form new group to protect electric grid from cyberattacks. NextGov, May 1, 2010. http://www.nextgov.com/cybersecurity/2010/01/energy-set-to-form-new-group-to-protect-electric-grid-from-cyberattacks/45641/.
35. Barret L. U.S. reviewing cyber threat to power grid. Internet News, September 15, 2009, www.internetnews.com/security/article.php/3839241.
36. Hamilton T. Smart grid saves power, but can it thwart hackers? TheStar.com, August 3, 2009. http://www.thestar.com/printArticle/675453.
37. Gross G. Lawmakers: Electric utilities ignore cyber warnings. *Computerworld*, July 21, 2009. http://www.computerworld.com/s/article/print/9135753/Lawmakers_Electric_utilities_ignore_cyber_warnings.

38. Gorman S. Electricity industry to scan grid for spies. *Wall Street Journal*, June 18, 2009.
39. Miller S. C. Our infrastructure in their crosshairs. *The News & Observer*, May 12, 2009.
40. Smart grid offers savings, vulnerabilities. *HS Daily Wire*, April 30, 2009. http://www.homelandsecuritynewswire.com/smart-grid-offers-savings-vulnerabilities.
41. Critics: Cybersecurity standards for grid do not go far enough. *HS Daily Wire*, May 1, 2009. http://www.homelandsecuritynewswire.com/critics-cybersecurity-standards-grid-do-not-go-far-enough.
42. Sarwate A. Hot or not: SCADA security is hot. *SC Magazine US*, April 23, 2009.
43. Mills E. Just how vulnerable is the electrical grid? CNET.News, April 10, 2009. http://www.cnet.com/news/just-how-vulnerable-is-the-electrical-grid/.
44. Meserve J. "Smart Grid" may be vulnerable to hackers. CNN. com. http://www.cnn.com/2009/TECH/03/20/smartgrid.vulnerability/index.html.
45. Madrigal A. Report: A smart grid is a hackable grid. TheAtlantic.com, October 7, 2010. http://www.theatlantic.com/technology/archive/2010/10/report-a-smart-grid-is-a-hackable-grid/64231/.
46. Half of critical infrastructure providers have experienced perceived politically motivated cyber attack. *Transmission & Distribution World*, October 6, 2010. http://tdworld.com/smart-grid/half-critical-infrastructure-providers-have-experienced-perceived-politically-motivated-c.
47. Preventing catastrophic impacts from adverse cyber-physical events. *CISW-SG 2010 Smart Grid Survivability Workshop*, Arlington, VA, October 13–14, 2010. http://www.wikicfp.com/cfp/servlet/event.showcfp?eventid=11133&copyownerid=8367.
48. Sonenklar C. Obama admin tabs $3 billion for smart grid. http://www.heatingoil.com/blog/obama-admin-tabs-3-billion-for-smart-grid-1028/.
49. Shneerson E. M. Digital relay protection. *Energoatomizdat*, 2007, 549pp.
50. Gurevich V. I. Whether the relay protection is safe? *Energy-Safety and Energy-Economy Magazine*, 2, 2010, 6–8.
51. Gurevich V. I. Reliability of microprocessor-based protective devices—Revisited. *Journal of Electrical Engineering*, 60 (5), 2009, 295–300.
52. Gurevich V. Sophistication of relay protection: Good intentions or the road to hell? *Energize*, January/February, 2010, 44–46.
53. Gurevich V. I. Sensational "discovery" in the relay protection. *Power and Industry of Russia*, 23–24, 2009.
54. Gurevich V. I. Logic in free flight. *PRO Electrichestvo*, 2, 2011, 28–31.
55. Gurevich V. Tests of microprocessor-based protection devices. *PRO Electricity*, 1, 2008, 41–43.
56. Gurevich V. The hazards of electromagnetic terrorism. *Public Utilities Fortnightly*, June, 2005, 84–86.
57. Gurevich V. I. Problems of electromagnetic impacts on digital protective relays. *Components and Technologies*, 2, 2010, 60–64.
58. Gurevich V. I. Problems of electromagnetic impacts on digital protective relays. *Components and Technologies*, 3, 2010, 91–96.
59. Gurevich V. I. Problems of electromagnetic impacts on digital protective relays. *Components and Technologies*, 4, 2010, 46–51.
60. Gurevich V. I. Stability of microprocessor relay protection and automation systems against intentional destructive electromagnetic impacts. *Electrical Engineering & Electromechanics*, 5 (Part I), 2011; 6 (Part II), 2011.

61. Gurevich V. I. Cyber weapon against of power industry. *Energize*, 10, 2011.
62. Lyamets U. Y., Kerzhaev D. V., Nudelman G. S., Romanov U. V., Multidimensional relay protection. Abstracts of *the Second International Scientific-Technical Conference on Modern Trends in Development of Power System Protection and Automation*, Moscow, Russia, September 7–10, 2009.
63. Bulychev A., Nudelman G. Relay protection. Improvement through proactive functions. *News of Electrotechnics*, 4 (58), 2009.
64. Gurevich V. I. Sensational "discovery" in relay protection. *Energy and Industry of Russia*, 23–24 (139–140), December 2009.
65. CIA: Hacking electrical grids is possible. CNews.ru: Newsline. January 24, 2008. http://www.cnews.ru/news/line/index.shtml?2008/01/24/285018.
66. Gurevich V. I. Electromagnetic terrorism—The new reality of the 21st century. *The World of Technics and Technologies*, 12, 2005, 14–15.
67. Daamen D. Avant-garde terrorism. *Intentional Electro Magnetic Interference*, University of Twente, Enschede, the Netherlands, 2002, 23pp.
68. Report of the commission to assess the threat to the United States from electromagnetic pulse (EMP) attack. http://www.empcommission.org/docs/empc_exec_rpt.pdf.
69. Gannota A. The object of attacks—Electronics. *Independent Military Review*, 13, 2001.
70. Loborev V., Parfionov U., Fortov V. Collapses of noiseless explosion. *Literary Gazette*, 5 (5865), February 6–12, 2002.
71. Pokrovsky V. Electromagnetic factor. *Independent Gazette*, October 08, 2003.
72. Bäckström M. Is intentional EMI a threat against the civilian society? *SAAB Communication*, 2006. http://www.researchgate.net/publication/228863349_The_threat_from_intentional_EMI_against_the_civil_technical_infrastructure.
73. Gurevich V. Reliability of microprocessor-based relay protection devices—Myths and reality. *Engineer IT*, 5, Part I: 2008, 55–59; 7, Part II: 2008, 56–60.
74. Gurevich V. I. Some performance and reliability estimations for microprocessor based protection devices. *Electric Power's News*, 5, 2009, 29–32.
75. Gurevich V. I. How to rebuild relaying? *Energize*, 4, 2010, 36–39.
76. Gurevich V. I. Criteria of estimation for a relaying: Whether it is necessary to complicate a situation? *Electric Power's News*, 6, 2009, 45–48.
77. Shalin A., Microprocessor-based relay protection: Analysis of the efficiency and reliability is required. *News of Electrical Engineering*, 2, 2006.
78. Problems of microprocessor-based relay protection. http://digital-relay-problems.tripod.com.
79. Fedosov A., Pusenkov E. Problems arising from the introduction of microprocessor technology in the emergency control systems. *Power Stations*, 12, 2009.
80. Belyaev A., Shirokov V., Emelyantsev A. Digital terminals of RPA. Adapting to Russian conditions. *News of Electrical Engineering*, 5, 2009.
81. Gurevich V. I. The standardization in the field of microprocessor protection relays is necessary. *Electric Power's News*, 2, 2011, 34–42.
82. Gurevich V. I. The new concept of digital protective relays design. *Serbian Journal of Electrical Engineering*, 7 (1), 2010, 143–151.
83. Gurevich V. I. Perspectives for hybrid technology in relay protection and automation. *Components and Technologies*, 10, 2011, 70–73.
84. Gurevich V. Hybrid reed-solid-state devices are a new generation of protective relays. *Serbian Journal of Electrical Engineering*, 4 (1), 2007, 85–94.

2

Natural Electromagnetic Affects on Digital Protective Relays

2.1 Electromagnetic Vulnerability of DPR

The problem of electromagnetic compatibility (EMC) of electronic equipment has arisen together with the advent of this kind of equipment, as it receives and transmits electromagnetic radiation. The problems resulted from both internal interferences of assemblies and external emission of various origins. For decades, the problems of EMC have been the prerogative of specialists in electronics, radio engineering, and communications. Suddenly, over the last 10–15 years, this problem has become critical for the power industry. Of course, rather high electromagnetic fields have always existed in electric power facilities. However, electromechanical devices that have been applied for decades in automation, control, and relay protection were not addressed to electromagnetic fields too much as there were no significant EMC problems encountered. The last two decades has showed a sharp changeover from electromechanical to digital protective relays (DPRs) and automation. Moreover, the changeover included both the construction of new substations and power plants and the replacing of old electromechanical protective relays (EMPRs) at the old substations, built in those days when nobody assumed using microprocessor technologies, with the up-to-date DPRs. The latter have proved to be very sensitive to electromagnetic interference (EMI) coming "out of thin air," penetrating through operating power circuits, voltage circuits, and current transformers (CTs). Some malfunctions of DPR were caused by mobile phones [1] and similar types of equipment. There have been other cases, such as malfunctions of DPRs at operating capacities of Mosenergo, Ochakovskaya, and Zubovskaya substations. The operating algorithm of protection was affected by lightning, excavation work nearby, electric welding, and other types of interference. The Lipetsk substation start-up was postponed for 6 months due to faults of DPRs while they spent nearly US$ 1.5 million for the DPR. As a result, the substation was commissioned using a set of conventional defenses [2]. In practice, the shorting on the 110 kV side can cause protection failures on the 330 kV side, and interference during switching of the same voltage rating penetrated inputs (through the auxiliary circuits) of the relay protection apparatus operating under the other voltage rating [3].

According to Mosenergo, faults due to improper operation of relay protection amount to 10% out of total number of malfunctions and basically relate only to microelectronic-based and microprocessor-based relays [4]. Such a high percentage of malfunctions due to insufficient EMC results from the fact that the DPR interference sensitivity is much higher than that of traditional electromechanical protection. For example, according to Ref. [4] when electromechanical relay operation can be affected by the energy of 10^{-3} J, the energy of only 10^{-7} J causes the malfunctions of the microchips. The difference is about 4 orders of magnitude, or 10,000 times.

The level of damage depends on the insensitivity of each circuit component and the energy of the powerful interference as a whole, which can be absorbed into the circuit without the appearance of any defect or failure. For example, although the switching noise caused by the inductive load with an amplitude of 500 V is a twofold voltage surge, it is unlikely to lead to a failure of an electromagnetic relay with a 230 V AC coil due to its insensitivity to this kind of interference and its short duration (it lasts only several microseconds). The situation is different if the chip is powered from a 5 V DC source. Impulse interference with an amplitude of 500 V is a hundredfold higher than the supply voltage of the electronic component and leads to the inevitable failure and the subsequent destruction of the device. Surge resistance of the chips is several orders of magnitude lower than that of the electromagnetic relays [5]. Surge resulting from lightning and switching in power plants can damage or destroy both electronic devices and complete systems. Long-term statistics confirms that the number of such incidents of damage doubles every 3–4 years [5]. This statistic is in agreement with the so-called Moore's law [6]. In 1965, Moore showed that the number of semiconductor components in microchips doubles roughly every 2 years and this trend remains valid for many years. If some 10 years ago the so-called transistor–transistor logic (TTL) chip contained 10–20 elements per square millimeter and had a typical supply voltage of 5 V, now the popular chip can contain nearly a hundred complementary metal–oxide–semiconductor (CMOS) transistors on every square millimeter of the surface and has a supply voltage of only 1.2 V (Figure 2.1).

The up-to-date solid-state technologies, for example, silicon on sapphire (SOS), raise the number up to 500 elements per square millimeter of the surface [7]. It is obvious that such chips require even lower supply voltages and it is even more obvious that such improved microelectronics integrity reduces insensitivity of its components to high-voltage surge due to the diminishing thickness of insulating layers and reduced operating voltage of semiconductor elements.

This means that a faulty substation DPR would work properly during laboratory testing and make it impossible to determine the reason of the

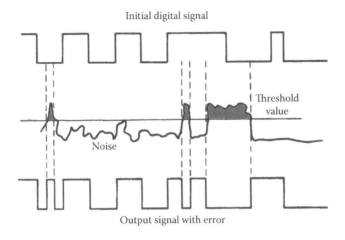

FIGURE 2.1
Impact of the low-energy interference onto digital devices.

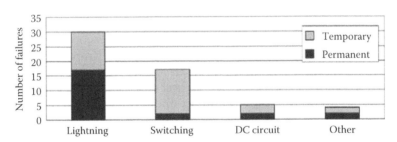

FIGURE 2.2
Statistics of Japanese manufacturers about destruction of DPR due to electromagnetic impact.

fault at the substation. Statistics gathered by the representatives of major Japanese manufacturers of DPR points clearly to this feature of the DPR (Figure 2.2) [8].

As seen from the chart, short-term nonrepeated malfunctions (faults) of DPR prevail in most cases. This was also confirmed by another group of researchers [9]. According to their data, such faults account for almost 70% out of the total number of DPR damage, while up to 80% of the faults occur in microchips.

According to Ref. [4] and practical experience of OJSC "Mosenergo" (Russia), there have been many cases of DPR failures caused by EMI. The most obvious was the inclusion of Siemens relay protection devices at TPP-12 of OJSC "Mosenergo" as per the design developed by the "Atomenergoproekt" Institute. The design didn't take the EMC requirements into account. As a result, within the period of August–December 1999 alone, there were more than 400 false information signals registered on discrete and analog inputs of DPR [4] due to interference impact. It should be considered that the cost of each DPR failure is 10 times higher than that of the electromechanical relay due to the high number of functions provided by each DPR.

2.2 Lighting Strikes

Lightning strikes are the most powerful sources of impulse effects on the equipment of power stations and substations. Nearly 50 lightning strikes hit the Earth's surface every second, and on the average, each square kilometer of the surface is stricken by the lightning six times per year.

The lightning voltage can be up to 100 million volts. Usually, construction standards for lightning rods assume that the lightning current is about 200 kA and its duration is about 1 ms; however, in reality, the lightning current rarely exceeds 20–30 kA. The temperature of the channel during the main discharge can exceed 25,000°C. The length of the lightning channel can be from 1 to 10 km, and its diameter can be several centimeters. When lightning strikes the lightning rod, the current (in the form of bell-shaped pulse [Figure 2.3]) flows into the ground and spreads in all directions up to dozens and even hundreds of meters away resulting in voltage drops due to ground resistance. Since the soil layers closest to the current entrance point have the largest resistance, they also show the highest voltage. The farther away from the point, the lower the current resistance and voltage (Figure 2.3).

In order to reduce the potential induced upon the flow of lightning current into the ground, residential and industrial buildings should be equipped with a large enough metal grid embedded under the foundations. However, the resistance of such grounding systems is still far from zero (Figure 2.3), and even the residual pulse potentials induced in the grounding system and penetrating through the cables to the inputs of electronic equipment are dangerous. Interference of this kind is called "conductive." Besides this interference, high-current impulses flowing through the lightning rod generate disturbing electromagnetic fields that affect all conductors located nearby. This is the so-called "inductive" effect. There are also capacitive pickups, when short (i.e., high frequency) surge voltage pulses from high-voltage power lines enter the low-voltage circuits through capacitive couplings of the transformer windings.

During the propagation, the interference transforms many times from one kind into another, so such classification is rather tentative, especially when it comes to high-frequency processes (lightning discharge current impulse with rather steep fronts of 8 and 20 ms [Figure 2.3] can be considered as such high-frequency process). Therefore, in order to make an accurate analysis of the current spreading into the ground through the grounding device, it is required to consider both of these components. Moreover, the interference,

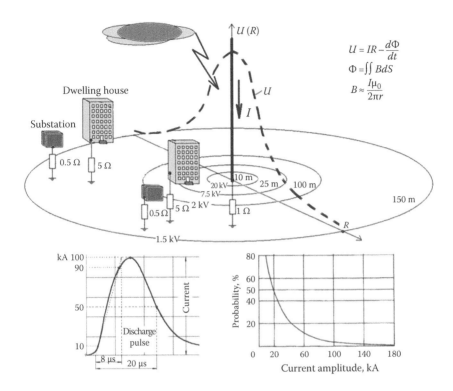

FIGURE 2.3
Processes occurring in the lightning rod upon lightning strike.

which entered electronic equipment through electromagnetic fields or conductors, undergoes multiple transformations immediately inside the equipment due to parasitic capacitive and inductive couplings between elements
or assemblies of the equipment. Thus, the high-frequency component of
interference can propagate into hardware, bypassing the filters and protective elements. There is also another way for lightning discharge interference
penetration: the current flowing through the grounded metal DPR casing
and grounded shields of numerous connecting cables. Therefore, it is very
difficult to provide a proper level of EMI protection. It is particularly difficult to realize at old substations where grounding systems were designed
for electromechanical protection, which is much more stable to electromagnetic impact than microprocessor-based one. Considering that dangerous
potential peaks in grounding circuits occur not only upon lightning strikes
but also upon emergency shorting in power mains, the problem becomes
even more complicated. In some cases, grounding rings of power equipment
are separated from those of electronic equipment to prevent such potential
peaks in electronic equipment circuits. However, such separation is not possible to arrange at existing substations. It is our opinion that the only way to

avoid the influence of high EMI on DPRs is to develop a complex solution. This solution should include

- Using DPR only at substations, which were designed and built considering the most up-to-date EMC requirements allowing the installation of the highly sensitive electronic equipment
- Improving the design of DPR
- Installing DPR in metal cabinets providing electronic equipment protection and equipped with filters on all cables entering the cabinet

2.3 Switching Processes and Electromagnetic Fields Generated by Operating Equipment

Switching processes and electromagnetic fields generated by operating electrical equipment are the second most influential source of impulse interference affecting the DPR under normal operating conditions.

The usual sources of switching noise in power industry are high-voltage switches and breakers, low-voltage relays and contactors, and controlled capacitor banks. Powerful electric drive frequency converters, corona discharge, and electrospark technologies generate electromagnetic emissions, which are dangerous for electronic equipment. Thus, the routes of interference penetration into DPR may be very different: from direct induced pickups at low-voltage conductors and substation secondary circuit cables (Figure 2.4) to surge and high-frequency overvoltage

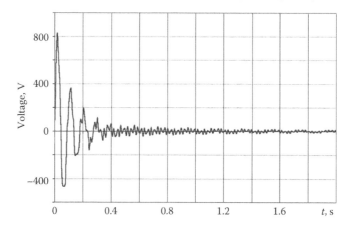

FIGURE 2.4

Voltage relative to the ground induced in low-voltage control cable upon switching process in a high-voltage circuit. (Data from Sowa, A.W. and Wiater, J., Overvoltages in protective and control circuits due to switching transient in high voltage substation, Electrical Department, Bialystok Technical University, Bialystok, Poland.)

occurring in the secondary windings of CT and voltage transformer (VT) (Figures 2.5 through 2.7).

The less the duration of arc's burning upon disconnection of the high-voltage circuit with switching apparatus, the greater the amplitude of induced voltage surge in the secondary circuits. Therefore, the highest overvoltage is generated by vacuum circuit breakers (CBs), then go SF_6 switch and oil switch and then a series of air CBs. This explains the different amount of DPR damage resulted from the operation of CBs and disconnectors insulated with SF_6 gas and air (Figure 2.8). It should be noted that the high-voltage interference

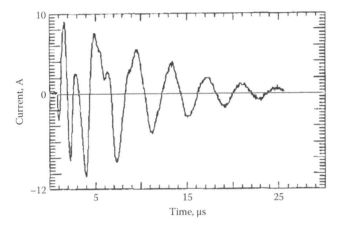

FIGURE 2.5
Current induced in the secondary winding of the CT upon switching the air CB with 500 kV voltage. (Data from Wiggins, C.M. et al., *IEEE Trans. Power Deliv.*, 9(4), 1869, 1994.)

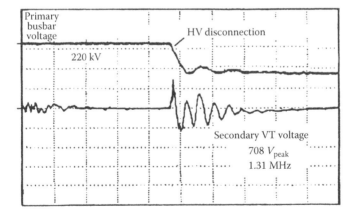

FIGURE 2.6
Pulse switching overvoltage in the secondary circuit of VT upon operating 220 kV air isolators. (Data from Carsimanovic, S. et al., Current switching with high voltage air disconnector, *International Conference on Power Systems Transients (IPST'05)*, Montreal, Quebec, Canada, June 19–23, 2005, Paper no. 229.)

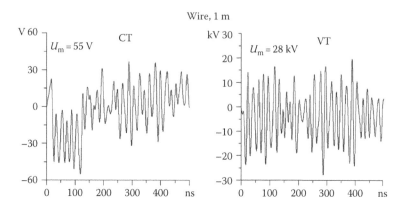

FIGURE 2.7
High-frequency noise in a 1 m long piece of conductor, connected to secondary circuits of CT and VT upon switching processes in SF$_6$ gas at 245 kV. (Data from Rao, M.M. et al., *IEEE Trans. Power Deliv.*, 22(3), 1504, 2007.)

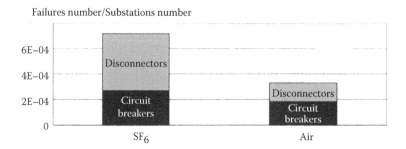

FIGURE 2.8
Comparative amount of DPR damages resulted from overvoltage upon the operation of CBs and disconnectors with SF$_6$ gas and air insulation. (Data from Matsumoto, T. et al., *IEEE Trans. Power Deliv.*, 21(1), 88, 2006.)

can be induced in the control cables by switching low-voltage circuits, particularly those that contain inductance. The nature of the switching transient process depends on many factors, and therefore the induced voltage can vary significantly even at the same substation. Since it is rather difficult to make theoretical calculations of such voltage surges, the easiest way to define them is to make direct measurements.

Significant voltage surges, transformed into the secondary circuits, occur also on capacitor bank switching (Figure 2.9).

An effective measure to avoid the induced voltage surge at the inputs of electronic equipment and its supply terminals is the wide use of elements with a nonlinear characteristic: gas-discharge arresters, varistors, special semiconductor elements based on voltage-regulator diode, etc., switched in parallel with the protected object (e.g., in parallel to the DPR input) and

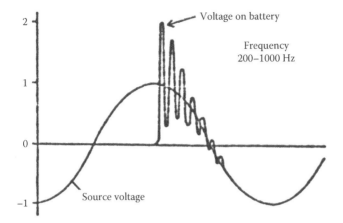

FIGURE 2.9
Overvoltage occurring upon capacitor switching.

between each terminal of the object and the ground. So far, the best results are obtained from resistors with a nonlinear characteristic, made from compacted ZnO powder (at least from silicon carbide, barium titanate, and other materials), such as the widely spread varistors.

Today, there are quite a few of them produced: without casing and in various casings with different auxiliary elements (fuses, alarm flags, etc.). Varistors must be properly selected. Unfortunately, wrong varistors are often installed even in equipment manufactured by leading world producers; thus, they provide no protection. Since the varistor current–voltage characteristic (IVC) is far from ideal (Figure 2.10), it is a problem how to choose it properly.

On the one hand, the varistor should not transmit current of more than 1 mA (default value for modern varistors of Western production) at maximum operating voltage (otherwise, it just overheats and burns). On the other hand, its "actuation" voltage, referred to as clamping voltage, should be much lower than that which the voltage of the protected equipment's electronic components (otherwise, these components will "protect" varistor) withstands. Since characteristics of IVC varistors do not allow them to fulfill these conditions, the maximum voltage that electronic components designed to operate in 220 V mains can withstand should not be less than 1000 V. However, such electronic components are more expensive than low-voltage ones, and their other characteristics are worse. For example, 1000–1200 V transistors have significantly lower gain factor and much higher open state voltage drops than the 400–500 V transistors of the same type. Therefore, the transistors with a maximum voltage of 500 V that they can withstand are often used in 220–250 V circuits of DPR power supply sources, emergency recorders, and other electronic equipment of the world's leading manufacturers. Varistors cannot protect the electronic components at such ratios of the operating and maximum withstanding voltage.

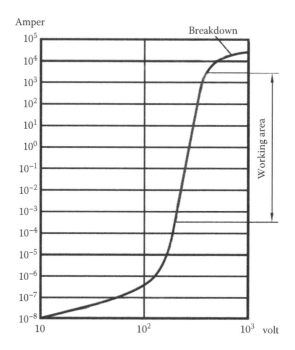

FIGURE 2.10
Typical IVC of zinc oxide varistors (ZOVs).

2.4 Issues with Control Cable Shielding

In order to protect control cables from induced voltages, they should be shielded and properly installed as far from the lightning rods and power cables as possible or in cable trays. There are several types of such trays: plastic trays with aluminum inserts, metalized plastic trays, and aluminum trays.

In general, the efficiency of metal screen (i.e., the degree of electromagnetic field attenuation) results from two properties: absorption of energy during the passage of electromagnetic waves through a conducting medium and the reflection of the wave at the interface between two mediums. Both of these phenomena depend on the electromagnetic wave frequency and the shielding material. The best electromagnetic absorption is provided by ferromagnetic materials (iron, permendur, permalloy), and the best electromagnetic wave reflection is provided by diamagnetic materials (copper, aluminum). The effectiveness of ferromagnetic material shielding decreases subsequent to the increase in field density due to saturation, while the effectiveness of diamagnetic shields decreases in proportion to frequency increase due to the increase in resistance. For a number of technical and economic reasons, copper mesh (braided) screens and aluminum profiles became the most popular.

Since the depth of electromagnetic wave penetration into metal is in an inverse ratio to the wave frequency, it is obvious that the thicker the metal shielding is, the better it will attenuate the electromagnetic field within a wide frequency range. For example, if 0.6 mm copper screen provides effective shielding at 500 kHz, you'd need a copper screen of about 6 cm thick for 50 Hz (for a ferromagnetic shielding, the thickness can be 5 mm).

Based on the foregoing, it is obvious that plastic metalized trays have the lowest shielding effect, while they are widely used for shielding control cables. Such construction becomes effective only at frequencies exceeding 600 MHz. At frequencies below 200 MHz, it does not work at all. Usually control cable pickups at substations have much lower frequency than the specified 200 MHz, so the use of plastic metalized trays is fully senseless. However, aluminum trays and copper braid on the cables are still able to reduce induced voltage most of the time; that's why they are more widely used. The greatest attenuation of the pickups in a wide frequency range can be provided by installing control cables in steel water pipes.

For successful operation of shielding, it is necessary to ensure that the induced charge flows down into the ground. Ideally, the potential along the entire length of the shielding should be equal to the ground potential, so sometimes shielding of very sensitive high-frequency electronic circuits is equipped with multiple grounding per every 0.2λ (λ—length of the electromagnetic field wave). At substations, shielded cables can be installed in parallel to the potential equalizing copper bus, grounded on both sides. However, the shielding is grounded more frequently on one or both sides (Figure 2.11).

Often protection engineers say that one-side grounding is sufficient for control cable shielding. Apparently, this resulted from such known measures as one-side grounding of current paths and one-side grounding of high-voltage cable shielding. Sometimes these two measures are used for grounding of control cable shielding regardless of the fact that grounding in these examples ensures electrical safety, but not protection from interference.

In fact, one-side grounding of the control cable shielding is effective only for capacitive interference (Figure 2.12) (the so-called electrostatic protection), but not for inductive interferences (attenuation factor $k = 1$), since the shielding does not provide interference current closure circuits.

Two-sided shield grounding provides additional circuit (shield) with much lower signal impedance for high frequency than the ground. As a result, the operating signal is divided into two parts, one of which (low frequency) returns through the ground, while the other (high frequency) goes through the cable shielding. Thus, for high-frequency component, the shielding current is equal to the central cable's core current, going in the opposite direction, and is compensated by electromagnetic coupling between the shielding and the central cable's core. This also provides protection against high-frequency emissions from the central cable's core to the external space (i.e., to the neighboring cables) with the interference attenuation factor $k = 3$–20. This system is also effective with the external electromagnetic effects to the shielding, when

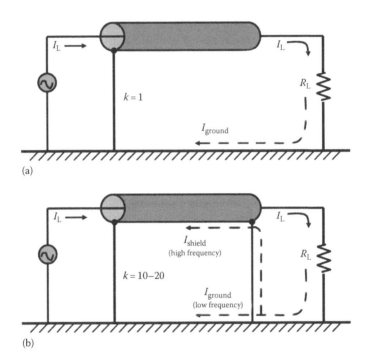

(a)

(b)

FIGURE 2.11
(a) Operation of a shield grounded on one side and (b) on both sides.

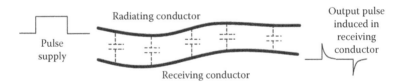

FIGURE 2.12
Pulse interferences through capacitive coupling between conductors.

a high-frequency signal induced to the shielding is closed to the ground. When connecting the shielding to the ground bus, you should avoid both wrapping the connecting cable over the shielding and its coiling in order to reduce its length between the shielding and the ground bus. Each additional coil of the cable increases high-frequency impedance of the grounding system and significantly reduces its effectiveness.

Sometimes the sources of high-power interference at substations are unexpected and not obvious. For example, one Russian substation experienced false trips of one of the high-voltage CBs upon commands to the trip coil of the other CB. Control cables going to the trip coils of both CBs were not shielded and were installed in the same tray for approximately 25 m. Voltage oscilloscope studies showed that the pulses with an amplitude of 500–728 V

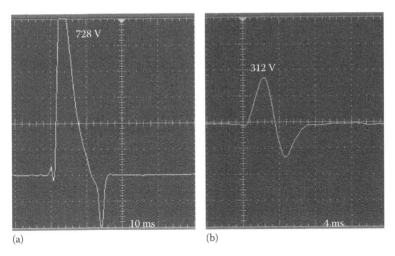

FIGURE 2.13
Oscillograms of induced interference pulse from one control cable to another: (a) nonshielded cables, (b) one cable shielded and grounded on both sides.

may be induced to the trip coil of the CB1 by supplying 220 V control voltage to the trip coil of CB2. Sometimes the duration of such induced pulse leads to a false trip of the CB (Figure 2.13a). The appearance of such high-power pulse interference in the control circuits is confusing and even puzzling. Everything becomes clear if one recalls that the CB trip coil is equipped with a ferromagnetic core and has a relatively large inductance, while the CB is equipped with an auxiliary contact, breaking the current in the coil upon CB trip. It is well known that the energy released upon the breaking of current circuit with inductance can be quite significant. After the control cable shielding at one of the CBs is grounded on both sides, the power of the induced interference pulse at the second cable was significantly lower (Figure 2.13b) and the false trips of the second CB were completely eliminated.

Two-side grounding may cause problems only if significant AC currents flow through the central conductor (typically, industrial frequency currents), generating high induced currents and heating the shielding. As a result, we should use larger wires (to reduce the heating of wire insulation) or ground one end of the shielding through the capacitor. The resistance of the capacitor is very high for the industrial frequencies, but it is very low for high-frequency interference.

In some cases, significant interference pulse currents can flow through a shield grounded on both sides, generating interferences into the central core. This can result, for example, from the high lightning strike current flowing through the grounding elements located near control cables or occurring under the near short-circuit current (Figure 2.14). As shown in Ref. [14], the peak voltage of interference at the central cable's core can reach up to 8.2 kV if the lightning's current is about 100 kA in the grounding conductor, even if the cable shielding is grounded on both sides. This is significantly higher

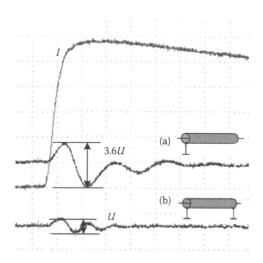

FIGURE 2.14
Voltage pickups at control cables with one-side (a) and two-side, (b) groundings of the shielding during the flow of lightning's current pulse (I) through the grounding conductor.

than the DPR's resistance level. In these cases, we must either change the control cable's route (install them further from the power switching devices, lightning rods, and arrester) or reduce the potential difference between the grounded ends of the cable shielding during powerful pulse interference. In order to do this, copper bus should be installed along cables.

This so-called "potential equalizing bus" should be well grounded on both ends. The copper bus impedance at high frequencies is much lower than the ground impedance (and even the shielding impedance), and so the largest amount of high-frequency current of pulse interference will flow through this bus, rather than through the shielding. These measures, combined with the aforementioned varistors, will ensure reliable protection of DPR. Certainly, these measures will be the most effective if adopted at the stage of design and construction of new substations, instead of "patching up" old substations.

The filter made of ferromagnetic (ferrite) ring or cylinder, which is mounted on the cable, has the opposite effect (Figure 2.15).

The impedance of the coil, which consists of one or two turns of control cable running through a ferrite ring, is very small for low-frequency operating signals and for the AC of industrial frequency, while it is very high for high-frequency (pulse) signals within a certain frequency range, which depends on the number of turns and the material of the ring. As a result, pulse and high-frequency interference, penetrating the cable, will be substantially reduced. Such filters can be widely used for relay protection in supply circuits, logic signals' transmission circuits, and the secondary circuits of CT and VT (Figure 2.16).

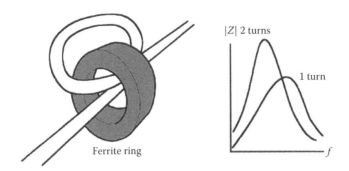

FIGURE 2.15
Ferrite ring filter and its frequency response.

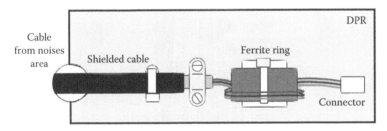

FIGURE 2.16
Installation of a ferrite ring filter on the control cable entering the DPR.

The effect of differential interference on the unshielded control cable (arising due to the difference between potentials induced on the direct and reverse wires) can be reduced by arranging conductors in such a way that the induced interferences will be similar, but opposite in voltage sign. This is achieved by twisting two AC conductors (direct and reverse). This method is effective if frequencies do not exceed 5 kHz, while its effectiveness depends on the uniformity and density of the wire twisting.

2.5 Distortion of Signals in the Current Transformer Circuits

To discuss the distortions introduced by CTs in the signal controlled by protective relays, let's consider a typical design and characteristics of traditional CT. High-voltage CTs, designed for installation in electrical networks (Figure 2.17), are equipped with a primary conductor (coil) 7 connected to the high-voltage line gap and several independent secondary windings with own core (kern) 8. The basic insulation in such transformers is performed by multiple layers of special paper strips wrapped around the

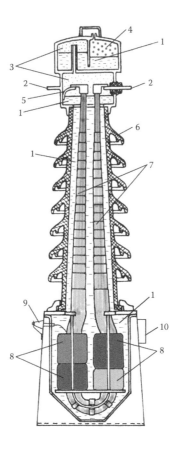

FIGURE 2.17
One of the common designs of high CT (160 kV) with paper–oil insulation: (1) oil, (2) copper plates to be included into the controlled current circuit, (3) containers with nitrogen, (4) air, (5) internal jumper connecting one of the output terminals with the upper reservoir's metal casing, (6) ceramic insulator, (7) isolated primary conductor (primary coil), (8) four independent secondary windings with own magnetic cores, (9) insulator on the outer wall of the container with the output terminal fixed to the container, (10) junction box with the secondary winding outputs.

primary coil and interleaved with thin aluminum foil poured with liquid transformer oil (Figure 2.17).

Since all the secondary windings are completely independent of each other and have separate cores, they should have been called the internal CTs. As a rule, these transformers have different electrical and magnetic characteristics, power, precision, etc. Low-power high-precision windings are designed for connecting measuring instruments, while more powerful, but less accurate, are designed for connecting protective relays.

As any other technical device, CT shows certain losses. Due to the loss, the primary current is not completely transformed into the secondary circuit. These losses cause the CTs *current error*. In addition, there is a phase shift in

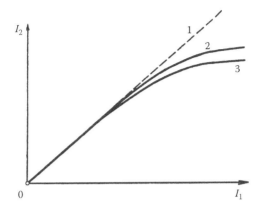

FIGURE 2.18
The relation of the secondary current (I_2) to the primary (I_1) in the CTs: (1) the ideal curve, (2) the real curve for the nominal load of the secondary circuit $Z_{2\,nom}$, (3) the real curve for the heavy load of the secondary circuit $Z_2 > Z_{2\,nom}$.

the secondary circuit in relation to the primary current, which causes the CT *angular error*. Losses in CT heavily depend on the magnetic circuit condition.

A direct proportion between primary and secondary currents remains, until iron in the magnet core is saturated. An increase of primary current results in an increase in the magnet core iron saturation, and the corresponding characteristic begins to bend (Figure 2.18). Moreover, the increase in the CT secondary circuit load (i.e., increase in load impedance) is followed by an increase in the characteristic's bending (since the effect of demagnetizing of the secondary current decreases).

It must be considered that these characteristics (Figure 2.19) are obtained during artificial CT testing and do not reflect the real relation between current and voltage under normal operating conditions of the CT. However, such artificial testing allows detecting various CT faults, and thus such CT characteristics are measured almost always upon the start-up of new equipment or during routine checks.

Measuring CTs operate within the rated current limits on the straight section of the characteristic providing high precision of measurements. Measuring CTs are available with accuracy classes of 0.2, 0.5, 1, 3, and 5 (the class number corresponds to the error in %). Protection relay CTs operate in emergency mode, under currents far exceeding the rated level, that is, on the magnetization curve arc, and their secondary current may be distorted (Figure 2.20). Therefore, the CT classes of relay protection include primary current limit ratio related to its rated value at which the error still exists. For example, the 5P20 indication means that the error of the CT does not exceed 5% at primary currents exceeding the rated value up to 20 times.

Under otherwise equal conditions, the power of the load connected to the CT's secondary circuit should not exceed the rated CT power in order to

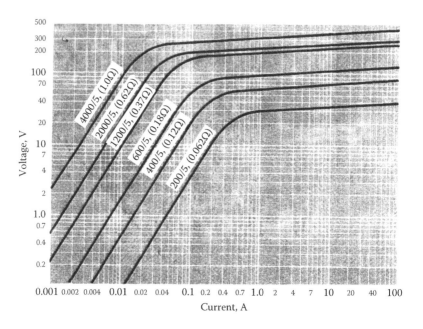

FIGURE 2.19
Actual IVCs of the CTs with different transformation coefficients represented in the manufacturer's documentation.

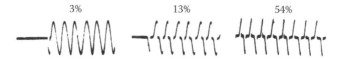

FIGURE 2.20
The form of the CT's secondary current under emergency mode (overcurrents). Error is shown in percents.

maintain the defined error. For a given nominal current, for example, 5 A, the load power is determined by its resistance:

$$P = Z_2 \times I_2^2,$$

where
Z_2 is the load resistance
I_2 is the secondary current

Therefore, the lower the resistance of the external circuit connected to the CT (i.e., relays), the lower the CT load degree and the smaller its error. The nature of the load also has a significant effect on the CT error as an increase in the inductive component of the load leads to the increase in current error and to the reduction of the phase displacement.

It would seem that the aforementioned information about the CT operation and its main characteristics is basic and must be known by any protection engineer. Surprisingly, it was found that at some networks the relay protection was connected to the measuring CT and at the same time there were complaints about the failures of relay protection. The research report presented at the *All-Russian Relay Protection Scientific Conference* under the auspices of UES Federal Grid Company [15] seriously revealed that since measuring cores of CTs produce highly distorted secondary current curve under the large primary current ratio it was not recommended to connect relay protection to such cores. The report also recommended to "further investigate this issue." We hope that there are very few such pseudo-protection engineers in Russia.

In fact, there were many CT distortion studies conducted. It is known that live connections of the power line lead to the occurrence of a transient process (Figure 2.21).

It is known that the current of the transient process $i(t)$ upon circuit connection depends on periodic and nonperiodic components. The latter depends on the circuit parameters. Under the purely active load, the nonperiodic component equals zero and the load current is absolutely sinusoidal (Figure 2.21a). The real power line has a certain inductance L, and therefore the nonperiodic component doesn't equal to zero.

This component shifts the current sine wave in relation to zero (Figure 2.21b) at the first moment after the line switches on (or upon the sharp increase in short-circuit current) and until the component damping moment. This shifted sine wave can be mathematically represented as the sum of two components: the usual sine wave I_{rms} and DC component (I_{DC}), which represents the DC current.

Transformers with nonlinear magnetization characteristics (i.e., common CTs) are maladjusted for operating under high nonperiodic component of

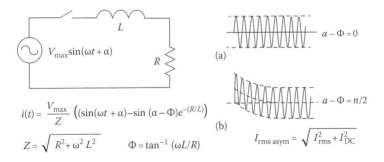

FIGURE 2.21
Transient process upon live connection of power lines: (a) for active load with the nonperiodic component equals zero; (b) for transformers with nonlinear magnetization characteristics.

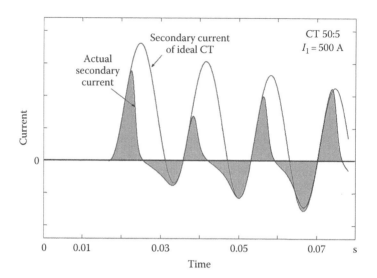

FIGURE 2.22
Distortion of the CT secondary winding's current under the constant component of the short-circuit current.

the transient process. The DC flowing through the CT's primary winding leads to a rapid saturation of the core and to a significant distortion of the periodic component's shape and, subsequently, that of the secondary current (Figure 2.22). CT's secondary current curve has the same shape when subjected to a magnetizing inrush current when the powerful supply transformer is turned on.

There are also significant distortions of the CT's secondary current, even if there is no DC component. At that, the saturation of the core occurs at very high ratio of the primary current (Figure 2.23).

At powerful electric energy systems with large short-circuit current ratio and significant time constant $L/R = 0.1$ s (Figure 2.21), the actual CT error can be very high (Figure 2.24) and the secondary current takes the shape of peaked pulses. Reduction of the CT's load and other measures have little influence on this error. The only measure that might reduce this error is a significant increase in wire section and mass of the CT magnet core iron. However, the calculations show that in this case the size of the CT can reach unacceptable values [16].

The easiest way to reduce the CT error in transient mode is to use CTs with primary rated current greater than the value obtained from the routine calculations. Thus, DPR setup must include real transformation ratio of the selected CT.

It is known [17] that the input circuits of DPR contain filter and analog-to-digital converters (ADCs) (Figure 2.25).

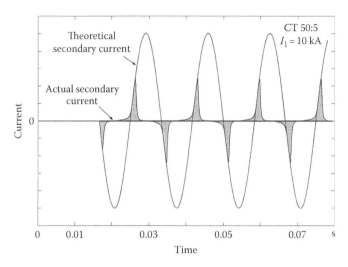

FIGURE 2.23
Distortion of current in the CT's secondary winding under the saturation at high primary current ratio.

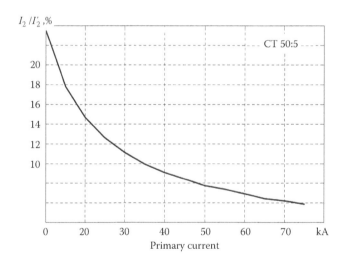

FIGURE 2.24
The amplitude of the real CT secondary current (I_2) in percentage of the ideal theoretical (I_2') value, depending on the primary current.

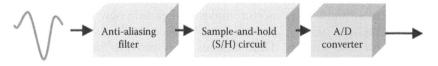

FIGURE 2.25
DPR analog signal input circuit (current, voltage).

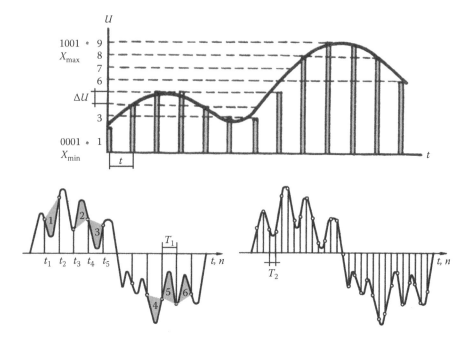

FIGURE 2.26
The operating principle of ADC and distortions that occur during transformation (sampling) of the signal.

The filter suppresses high-frequency signals, and since the distorted secondary current curve of CT contains significant high-frequency component, the filter reduces the amplitude of the signal. Then, the ADC discretely measures the periodic signal at specific time intervals t (Figure 2.26).

Because ADCs operate by sampling the input values at fixed intervals, it is clear that there is no way to determine the value of the input signal at intervals between such samplings (sectors 1–6 in Figure 2.26).

If the secondary current curve is highly distorted, the error increases significantly (Figure 2.27). As a result, the signal generated at the ADC output has almost nothing to do with the real current. Time–current characteristics of DPR are not maintained, sectors of remote protection are determined incorrectly, etc.

In order to compensate for this error, it is required to make the DPR input circuits significantly more complex, integrating new functional units (Figure 2.28) [18].

There are also specific DPR designs that provide fast computation of the second derivative of the current and use the obtained values for the threshold adjustment under the high nonperiodic component of the current [19].

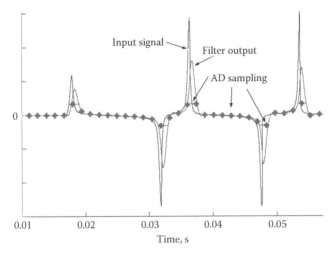

FIGURE 2.27
Transforming secondary CT current at DPR input circuits.

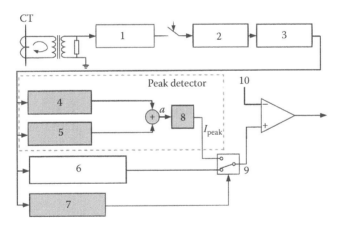

FIGURE 2.28
The structure of the DPR input circuits compensating the CT saturation error. (1) Filter, (2) ADC, (3) storage of the last 16 samplings, (4) detector of the maximum amplitude value, (5) detector of the minimum amplitude value, (6) peak meter for the basic frequency signal with traditional filter, (7) detector of the signal distorted by saturation, (8) divider by 2, (9) amplitude discriminator, and (10) signal of the current cutoff element.

2.6 Harmonics Impact in the Measured Current and Voltage on DPR

The harmonic distortion of the voltage and current entering the DPR input represents the components with higher frequencies compared to the main frequency. Therefore, their effect on DPR is similar to that described earlier and

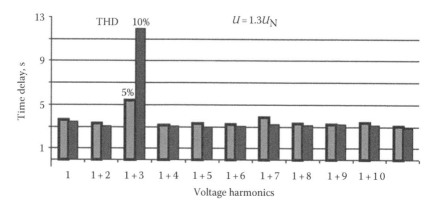

FIGURE 2.29
Effect of voltage harmonic distortion (5% and 10%) at the DPR analog input on the response time.

is also associated with filtering and sampling errors in the ADC [20]. Studies of several authors demonstrated that the third harmonic has the greatest influence on DPR. Among others, this harmonic usually has the greatest value of voltage and current that only exacerbates the situation. The study of microprocessor voltage relay with definite time delays [21] showed that this delay heavily depends on this harmonic distortion (Figure 2.29). Voltage control elements are used in many DPR types, such critical ones as remote line protections, which make this dependence on the harmonic distortion existence very dangerous.

As for the supply voltage harmonic distortion, its influence on DPR is much lower, as virtually all DPR types are supplied with switched power supply with a filter and rectifier installed at the input. Rectified and filtered voltage is converted into high-frequency voltage, changes in magnitude, is stabilized, and then is rectified and passes through the filter again [22].

All these conversions in switched power supplies reduce the effect of supply mains harmonic distortion on DPR to zero.

2.7 Quality of Voltage in the Supply Mains

In many cases, the malfunctions and even damages of microprocessors caused as a result of the impact of electromagnetic disturbances (*blackouts, noise, sags, spikes, surges*) from the power supply network on operation of the DPR.

2.7.1 Blackout

A blackout results in total loss of utility power.

- *Cause*: Blackouts are caused by excessive demand on the power network, lightning storms, ice on power lines, car accidents, construction equipment, earthquakes, and other catastrophes.

- *Effect*: Current work in random access memory (RAM) or cache is lost. It also results in a total loss of data stored on read only memory (ROM).

2.7.2 Noise

More technically referred to as EMI and radio-frequency interference (RFI), electrical noise disrupts the smooth sine wave one expects from utility power.

- *Cause*: Electrical noise is caused by many factors and phenomena, including lightning, load switching, generators, radio transmitters, and industrial equipment. It may be intermittent or chronic.
- *Effect*: Noise introduces malfunctions and errors into executable programs and data files.

2.7.3 Sag

Also known as brownouts, sags are short-term decreases in voltage levels. This is the most common power problem, accounting for 87% of all power disturbances according to a study by Bell Labs.

- *Cause*: Sags are usually caused by the start-up power demands of many electrical devices (including motors, compressors, elevators, and shop tools). Electric companies use sags to cope with extraordinary power demands. In a procedure known as rolling brownouts, the utility will systematically lower voltage levels in certain areas for hours or days at a time. Hot summer days, when air conditioning requirements are at their peak, will often prompt rolling brownouts.
- *Effect*: A sag can starve a microprocessor of the power it needs to function and can cause frozen keyboards and unexpected system crashes, which result in both lost or corrupted data. Sags also reduce the efficiency and life span of electrical equipment.

2.7.4 Spike

Also referred to as an impulse, a spike is an instantaneous, dramatic increase in voltage. A spike can enter electronic equipment through AC, network, serial, or communication lines and damage or destroy components.

- *Cause*: Spikes are typically caused by a nearby lightning strike. Spikes can also occur when utility power comes back online after having been knocked out in a storm or as the result of a car accident.
- *Effect*: Catastrophic damage to hardware occurs. Data will be lost.

2.7.5 Surge

A surge is a short-term increase in voltage, typically lasting at least 1/120 of a second.

- *Cause*: Surges result from the presence of high-powered electrical motors, such as air conditioners. When this equipment is switched off, the extra voltage is dissipated through the power line.
- *Effect*: Microprocessors and similar sensitive electronic devices are designed to receive power within a certain voltage range. Anything outside of expected peak and rms (considered the average voltage) levels will stress delicate components and cause premature failure.

Malfunctions and damages of DPR owing electromagnetic disturbances are described in literature. For example, mass malfunctions of microprocessor-based time relays occurred in nuclear power plants in the United States [23]. A review of these events indicated that the DPR failed as a result of voltage spikes that were generated by the auxiliary relay coil controlled by the DPR. The voltage spikes, also referred to as "inductive kicks," were generated when the DPR contacts interrupted the current to the auxiliary relay coil. These spikes then arced across the contacts of the output relay of DPR. This arcing, in conjunction with the inductance and wiring capacitance, generated fast electrical noise transients called "arc showering" (EMI). The peak voltage noise transient changed as a function of the breakdown voltage of the contact gap, which changed as the contacts moved apart and/or bounced. These noise transients caused the microprocessor in the DPR to fail.

According to IEC 61000-4 [24,25], voltage sags (sometimes referred to as "dips") are brief reductions in voltage (below $0.8U_N$), lasting from tens of milliseconds to 15 s (see Figure 2.30). As is known, the main reason for voltage sags in the 0.4 kV network of the substation auxiliary supplies is short circuits in external high-voltage grids. In manufacturing plants, such voltage sags are frequently associated with the working modes of the powerful electrical equipment, for example, with starting power motors. Voltage sags are an important criterion of the power quality.

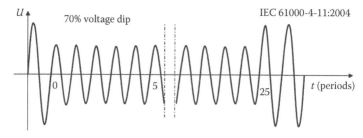

FIGURE 2.30
Example (from IEC 61000-4-11) of the 70% voltage sag during of 25 cycles (0.5 s).

The peculiarity of the low-voltage auxiliary AC network in power substations is that it does not contain devices that do not allow for short pauses in the power supply and almost all of the critical power consumers (relay protection, emergency modes recorders, communication system, signaling, and remote control) are fed, as usual, from power substation battery. At the same time, the power electronic systems with microprocessor controllers such as invertors, battery chargers, and power supplies are fed from the auxiliary AC network. Practical experience has shown that such devices do not "love" short voltage interruptions (50–200 ms) with subsequent voltage restoration. Sometimes such devices have time to hang through automatic changeover from main to spare power supply (transformer). Another problem of the power battery chargers with powerful input transformer is the high inrush current at sudden interruption and subsequent input voltage restoration that causes full charger disconnection by the electromagnetic releaser of the input CB. This state of affairs is considerably aggravated in some cases when even single voltage sags with durations of 100–200 ms provoke multiple pickups and releases of the electromagnetic contactor during the sags.

For increasing the reliability of AC network 0.4 kV in power substations, ordinarily two auxiliary transformers, feeding from different lines, are used. One of them connects to the 0.4 kV network permanently, and the other one automatically connects at voltage disappearance on the first transformer. Connecting and disconnecting AC network 0.4 kV to these transformers is affected by means of two powerful electromagnetic contactors on currents 200–400 A with AC control coils. These contactors are the major elements of the auxiliary network on which, in many respects, reliable work of all power substations depends.

As an object of research, the electromagnetic contactor 3TF54 type (Siemens) with switching capability 300 A has been taken (Figure 2.31), which is used at changeover in the auxiliary AC network 0.4 kV in power substations. During the research, the oscillograms of pickups and releases of the contactor have been recorded at the feeding of the control coil from the AC supply (Figures 2.31 and 2.32). The oscillogram in Figure 2.31 shows the presence of high starting (inrush) current caused by small impedance of the coil until the moment of the closing of the contactor's magnetic circuit. Oscillograms shown in Figure 2.32 allow determining the time pickups and time release of the contactor, that is, the reaction time of the contactor for voltage sags.

Analysis of the oscillograms has shown that full switch-on time, that is, the time interval between the moment that the voltage is applied to coil and the moment of main contacts' closings, is about 20 ms. And the full switch-off time, that is, the time interval between the disconnection of voltage on the coil and the moment of main contacts' openings, is 15–18 ms.

According to the data sheet, 10–30 ms for nominal voltage is applied before disconnection and 10–15 ms for voltage 0.8 of the nominal value. Such small time delays for such large contactors mean that during typical voltage

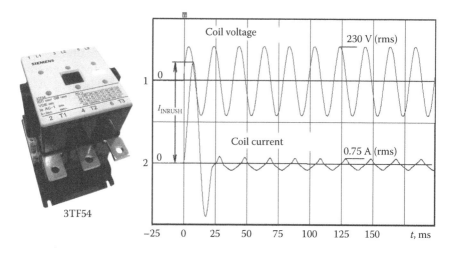

FIGURE 2.31
Electromagnetic AC contactor 3TF54 type (Siemens) and oscillograms for current and voltage on its coil at switch on.

disturbances with alternate voltage level sags and restorations, the contactor will have time to connect and disconnect the main power circuit several times. Moreover, as shown in Ref. [26], the reaction of the contactor during a 75% voltage sag is more serious than 100% since the releasing time for the first case is shorter on 40%–50% than in the second case and may be 10 ms even for large apparatus.

One more problem of the power AC contactor was discovered during the research stage. It was found that reducing voltage across the coil up to 150–135 V causes high vibration of the contactor magnetic system with a magnitude that is sufficient for closing and opening its main contacts. The same phenomenon arises when the AC voltage across coil increases from 0 up to 160–185 V. What this means is that a contactor working in such conditions, together with it fast acting, even at a single voltage sag (to level 135–150 V during 100–200 ms [Figure 2.33]), transforms in power generator multiple interruptions of the main voltage in substation auxiliary AC network. The same result appears when trying a connection of the contactor's coil to a power supply with a voltage of 150–170 V.

In view of the character of the loads fed from the auxiliary AC network in substations (power electronic equipment sensitive to short sags), the technical solution offered for contactors intended for use in manufacturing plants (retention of the contactor in a closed position at short sags) is not the correct technical solution for substation 0.4 kV network, because, through closed contactor's contacts, the short sags will affect sensitive equipment and provoke its disturbances.

In our opinion, the problem must be solved not by means of retention of the contactor in the closed position at voltage sags but by means of the rapid

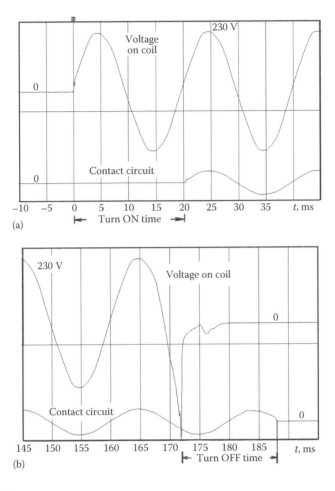

FIGURE 2.32
Oscillograms of contactor, type 3TF54, (a) switching on and (b) switching off.

(during 10–12 ms) disconnection of the contactor's coil at voltage levels drop-ping below 160 V and returning it to initial condition at voltage level resto-ration up to 185 V with time delay of 5–10 s. A single interruption of 5–10 s duration in the auxiliary 0.4 kV AC substation power network does not cause serious disturbances in the substation equipment due to a power battery feeding most of the important substation consumers. At the same time, such algorithm of the contactor's work may prevent serious disturbances and fail-ures in AC power electronic equipment.

For fast contactor disconnection at voltage network drops, most electronic relays available in the market are not suitable because their minimal reac-tion time is, usually, 100 ms. During a time interval of this duration, the contactor will make several disconnections and connections of the main power.

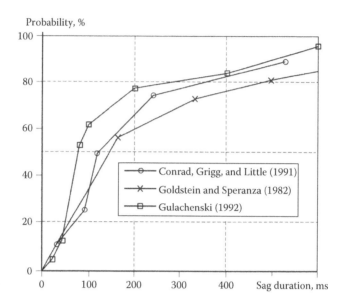

FIGURE 2.33
Sags duration in power AC network according to some researches.

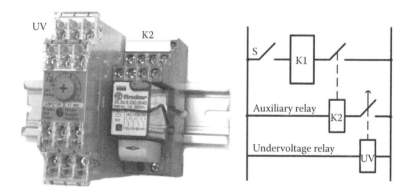

FIGURE 2.34
Device for fast forceful disconnection of the main contactor at voltage sag.

As a result of our research, only a few types of the devices suitable for contactor's control were found (Figure 2.34). One of them is an undervoltage relay combined with timer: Brownout Timer GBP2150 type, manufactured by Midland Jay (division of Midland Automation, United Kingdom). The reaction time of this device for voltage drops of 30% is 5 ms only. The release time after voltage restoration up to 80% can be adjusted to intervals of 1–10 s. Another good example is the Russian undervoltage relay PKH-1-3-15 type, manufactured by ZAO "Meander" (St. Petersburg). Such devices are ideal solutions for our purposes. For decreasing loads on the output contacts in these devices, an additional auxiliary fast electromagnetic relay, type 58.32.8.230 (Finder), with powerful output contacts and releasing time 3 ms is employed.

The problem of voltage dips in relay protection power supply circuits (AC, first of all) is as much important as voltage dips in the auxiliary power plants and substations supplying their own needs.

In the era of simple electromechanical relays, which did not require external power supplies (as DPR), the problem was limited to the necessity of supplying power to the trip coil of a CB during voltage dips. The problem was successfully solved 50 years ago using special units with reservoir capacitors, which supplied a necessary pulse of discharging current in the trip coil of a CB. The capacitor in this unit receives a constant charge from the AC circuit through a D1 diode (Figure 2.35).

In order to limit the capacitor's charging pulse current, which goes through D1 diode during connection of the noncharged capacitor to power supply, posistor TDR1 is used, while in order to protect the capacitor from spikes of overvoltage, a varistor M1 is used. Some manufacturers (e.g., Square D) add LED indicators associated with the power to their circuits as well as threshold elements on thyristors, which ensure full charging of the capacitor until it is discharged to the load and used during multiple charging cycles (Figure 2.36).

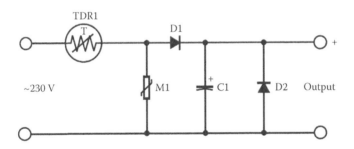

FIGURE 2.35
A typical layout of a capacitor unit, the purpose of which is to supply power to trip coil of a CB during voltage dips in the relay protection and auxiliary supply circuits.

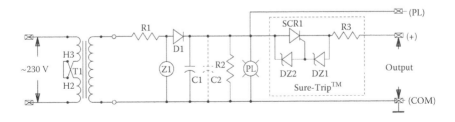

FIGURE 2.36
A complicated layout of a capacitor power unit Sure-Trip (Square D) with additional threshold element on a thyristor SCR1, LED PL, and Zener diode Z1.

For more complicated electromechanical protection, which includes many auxiliary relays in their internal circuits, power supply is necessary during voltage dips. The pulse accumulators of energy based on capacitors described earlier cannot supply the power of such relays during the time necessary for their actuation, especially if there is time delay relays.

As a result, complex power units appeared that supply power to relays simultaneously from a CT, a VT, and a capacitor (Figure 2.37).

Similar devices are manufactured by dozens of companies, including leading electrotechnical companies, such as General Electric, Siemens, ABB, and Alstom (Figure 2.38).

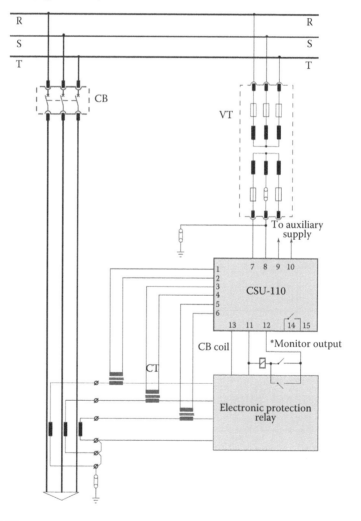

FIGURE 2.37
A connection diagram of a capacitor unit CSU-110 (Switching Systems Electronic Engineers) with a complex power supply from CTs, VTs, and the auxiliary power supply.

FIGURE 2.38
Complex power units of different manufacturers.

Internal self-discharge of large capacity electrolytic capacitors, which are used in capacitor power units, is limited in time, within which the capacitor saves energy sufficient for reliably tripping a CB. This is why built-in Ni-Cd baby batteries are used to charge capacitors in the most advanced models of capacitor blocks, when all external power supply options are lost. These baby batteries are constantly charging under normal regime, but when external power is lost, they supply power to a built-in low-power converter, which transforms the low voltage of the baby battery into high voltage, necessary to charge a capacitor.

The energy of baby batteries in these units (Figure 2.39) is enough to maintain a capacitor fully charged for 72 h when all types of external power supply are fully lost. Such compact devices are issued by many companies and allow keeping the capacitor charged for several days. Clearly, in such conditions, sufficient reliability of relay protection, even on an auxiliary AC, is provided. For this reason, the auxiliary AC is applied widely.

The situation began to change with the introduction DPRs and the mass replacement of electromechanical relays by them. To the many problems caused by this transition [27], one more problem was added. As is known, the internal switching-mode power supply, admitting use as auxiliary AC and DC voltages, has an overwhelming majority of DPRs. Therefore, at first sight, there should be no reasons to interfere with the use of an auxiliary AC voltage on substations with DPR. The problem arises when there is not

FIGURE 2.39
Capacitor power units with Ni-Cd baby batteries, maintaining capacitor's charge during 72 h.

enough power for normal operation of an overwhelming majority of DPR and only the presence of corresponding input signals (as for electromechanical relays) and also requires a feed from an auxiliary supply. How will the DPR behave at loss of this feed at failure mode when the hard work of the microprocessor and other internal elements is required? How will the complex relay protection (containing some of DPR, incorporated in the common system by means of the network communication when there are also losses of auxiliary feed) function? How will the DPR behave during voltage sags (brief reductions in voltage, typically lasting from a cycle to a second or so or tens of milliseconds to hundreds of milliseconds) during failure? We shall try to understand these questions.

The internal switching-mode power supply of the DPR contains, as a rule, a smoothing capacitor of rather large capacity, capable of supporting the function of the relay during a short time period. According to a research that has been led by the General Electric [28] for various types of DPRs, this time interval takes 30–100 ms (only for modern DPR of a new generation, the situation is otherwise [Table 2.1]). In view of that time of reaction, the DPR for emergency operation lays in the same interval and depends on that

TABLE 2.1

Power Interruptions Withstanding by Modern DPRs of Some Types

DPR Type and Manufacturer	Maximal Power Interruption Duration without Disturbances in Relay Functions	Minimal Level of the Voltage Supply Need for Proper Functioning of DPR with 230 V Nominal Voltage
SIPROTEC 7UT6135 Siemens	1.6	78
SIPROTEC 7UT6125 Siemens	1.6	36
SIPROTEC 7SJ8032 Siemens	3.8	44
T60 General Electric	—	80
P132 Areva	—	45

type of emergency mode; it is impossible to tell definitely whether protection will have sufficient time to work properly. At any rate, it is not possible to guarantee its reliable work. It is a specially problematic functioning of protection relays with the time delay, for example, the distance protection with several zones (steps of time delay, reaching up to 0.5–1.0 s and more). Also it is possible to only guess what will take place with the differential protection containing two remote complete sets of the relay, at loss of a feed of one of them only.

Voltage sags are the most common power disturbance. At a typical industrial site, it is not unusual to see several sags per year at the service entrance and far more at equipment terminals.

These voltage sags can have many causes, among which may be peaks of magnetization currents, most often at inclusion of power transformers. Recessions and the rises of voltage arising sometimes at failures and in transient modes are especially dangerous when coming successively with small intervals of time. The level and duration of sags depend on a number of external factors, such as capacity of the transformer, impedance of a power line, remoteness of the relay from the substation transformer, and the size of a cable through which feed circuits are executed. DPRs also have a wide interval of characteristics on allowable voltage reductions. As mentioned in Ref. [27], various types of DPRs keep working capability at auxiliary voltage reduction from the rated value of up to 70–180 V. Thus, DPR with a rated voltage of 240 V supposes a greater (in percentage terms) voltage reduction than devices with rated voltage 120 V. It is also known that any microprocessor device demands a long time from the moment of applying of a feed (auxiliary voltage) to full activation at normal mode. For a modern DPR with a built-in system of self-checking, this time can reach up to 30 s. It means that even after a short-term failure with auxiliary voltage (voltage sag) and subsequent restoring of voltage level, relay protection still will not function for a long time.

What is the solution to the problem offered by the experts [28] from General Electric? Fairly marking that existing capacitor trip devices obviously are not sufficient to feed DPR, as reserved energy in them has enough only for creation of a short duration pulse of a current and absolutely not enough to feed DPR, the author comes to the conclusion that it is necessary to use an uninterruptible power supply (UPS) for feeding the DPR in an emergency mode. The second recommendation of the author—to add an additional blocking element (a timer, e.g., or internal logic of DPR)—will prevent closing of the CB before the DPR completely becomes activated. Both recommendations are quite legitimate. Here, only the usage of UPS with a built-in battery is well known as a solution for the maintenance of a feed of crucial consumers in an emergency mode. This solution has obvious foibles and restrictions (both economic and technical). The use of blocking for switching on of the CBs can be a very useful idea that should be undoubtedly used; however, it does not always solve the problem as failures of voltage feeding connected to operation of the CB are always a possibility.

In our opinion, a more simple and reliable solution of the problem is use of a special capacitor with large capacity connected in parallel to the feed circuit of every DPR instead of UPS usage. High-quality capacitors with large capacity and rated voltage of 450–500 V are sold today by many companies under the price, approximately, €150–€200, and are not deficient; see Table 2.2.

Elementary calculation shows that when charged up to a 250 V, one 5000 μF capacitor is capable of feeding a load with consumption power 30–70 VA up

TABLE 2.2

Parameters of Capacitors with Large Capacity and Rated Voltage of 450–500 V

Capacity (μF)	Rated Voltage (V)	Dimensions (Diameter, Height) (mm)	Manufacturer and Capacitor Type
6,000	450	75 × 220	EVOX-RIFA PEH200YX460BQ
4,700	450	90 × 146	BHC AEROVOX ALS30A472QP450
10,000	450	90 × 220	EVOX-RIFA PEH200YZ510TM
4,000	500	76.2 × 142	Mallory DuraCap 002–3052
4,000	450	76.2 × 142	CST-ARWIN HES402G450X5L
6,900	500	76.2 × 220	CST-ARWIN CGH692T500X8L

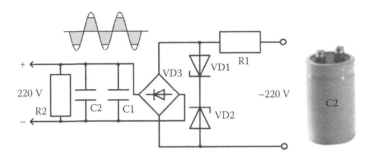

FIGURE 2.40
The device for reserve feed of DPR at emergency mode with AC auxiliary voltage.

to voltage decreasing to a minimum level of 150 V during 3–5 s, which is quite enough for operation of the DPR in the emergency mode.

Use of such capacitor for auxiliary voltage of 220 V AC requires, naturally, a rectifier and some more auxiliary elements (Figure 2.40).

In this device, a capacitor of large capacity is designated, such as C2. The C1 auxiliary—not electrolytic—capacitor with a capacity in some microfarads serves for smoothing pulsations on electrolytic capacitor C2. It is possible to include also in parallel to C1 one more nonelectrolytic capacitor with a capacity of some microfarad, for protection of C2 against the high-frequency harmonics contained in mains AC voltage. A resistor R1 (200–250 Ω) limits the charging current of C2 at a level near 1 A. The same resistor also limits pulse currents proceeding through back-to-back connected Zener diodes VD1 and VD2. A resistor R2 has high resistance and serves to accelerate the discharging capacitor up to a safe voltage at switching off of the auxiliary voltage. Zener diodes are intended for the maximal value voltage limits of capacitor C2 at a level of 240 V. Without such limitations on the device, output voltage would reach a value of more than 300 V due to the difference between rms and peak values of voltage. That is undesirable both for MPD and for C2.

The Zener diodes slice part of a voltage sinusoid in which amplitude exceeds 240 V, forming a voltage trapeze before rectifying. As powerful Zeners for rating voltage above 200 V are not at present on the market, it is necessary to use two series of connected Zeners with dissipation power of 10 W and rating voltage of 120 V, for each Zener (VD1, DD2, e.g., types 1N1810, 1N3008B, 1N2010, NTE 5223A).

As further research of this type of situation clarified, the problem of maintenance of reliable feed DPR is relevant not only for substations with AC auxiliary voltage but also for substations with DC voltage. Many situations where the main substation battery becomes switched off from the DC bus bars are known. In this case, nothing terrible occurs, as the voltage on the bus bar is supported by charger. However, if during this period an emergency mode occurs in a power network, the situation appears to be no better since use of an AC auxiliary voltage as charger feeds from the same AC network. Usually an electrolytic capacitor with some hundreds of microfarads for smoothing voltage pulsations is included on the charger output. Since not only many DPRs but also a set of other consumers are connected to charger output, it is abundantly clear that this capacity is not capable of supporting the necessary voltage level on the bus bars during the time required for proper operation of the DPRs. Our researches has shown that such high capacitance as 15,000 μF does not provide proper functioning of DPR at consumption from charger that reaches up to 5–10 A.

For maintenance of working capability of DPRs in these conditions, it is possible to use the same technical solution with the individual storage capacitor connected in parallel to each DPR feeding circuit. Now the design of the device will be much easier due to a cutout from the circuit diagram of Zeners and bridge rectifier (Figure 2.41). The resistor R (100 Ω) is necessary for limiting of the charging current of the capacitor at switching on auxiliary voltage with a fully discharged capacitor. Diode VD1 should be for a rated current of not less than 10 A. High-capability quick-blow fuse F (5 A/1500 A, 500 V) is intended for the protection of both the feeding circuit of DPR and the external DC circuit at damaging of the capacitor.

The prototype of such device with the capacitor 3700 μF (Figure 2.42) has shown excellent results at tests, with the various loadings that simulate DPRs of various types with different power consumptions at a nominal voltage of 240 V.

American Schweitzer Engineering Laboratories, Inc. (SEL) has changed their tack. It produces a complex power supply unit for the DPRs containing a small 48 V battery and charger (see Figure 2.43).

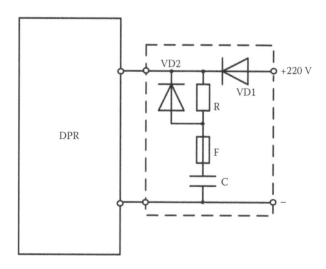

FIGURE 2.41
The device for reserve feed of DPR at emergency mode with DC auxiliary voltage.

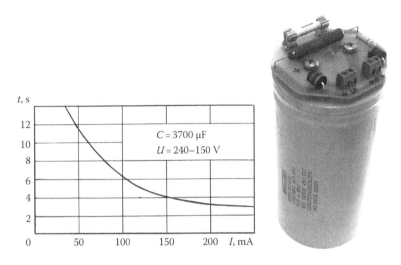

FIGURE 2.42
The prototype of the device for reserve feed of DPR.

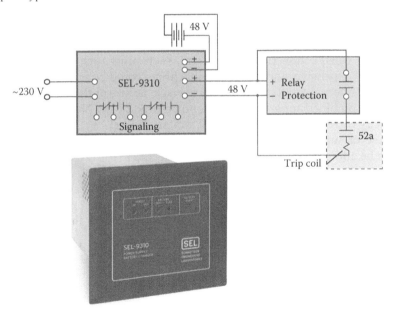

FIGURE 2.43
Complex power supply unit with charger and small 48 V battery manufactured by SEL.

Relay protection systems equipped with such devices can easily withstand prolonged voltage dips in auxiliary power supply or even long-lasting full blackouts.

One more variant of the solution of this problem for substations with DC auxiliary voltage is to not use an individual capacitor for each DPR, but rather a special "supercapacitor" capable of feeding a complete relay protection system set together with conjugate electronic equipment within several seconds.

Such supercapacitors can already be found in the market under brand names such as "supercapacitors," "ultracapacitors," "double-layer capacitors," and also "ionistors" (for Russian-speaking technical literature). There are electrochemical components intended for storage of electric energy. On specific capacity and speed of access to the reserved energy, they occupy an intermediate position between large electrolytic capacitors and standard accumulator batteries, differing both from one and the others in their principle of action, based on redistribution of charges in electrolyte and their concentration on the border between the electrode and electrolyte.

The capacity of modern supercapacitors reaches hundreds and even thousands of farads. Today, supercapacitors are produced by many Western companies (Maxwell Technologies, NessCap, Cooper Bussmann, EPCOS, etc.) and also some Russian enterprises (ESMA, ELIT, etc.); however, the rated voltage of one element does not exceed, as a rule, 2.3–2.7 V. For higher voltage, separate elements connecting among themselves in parallel and series as consistent units.

Unfortunately, supercapacitors are not so simply incorporated among themselves as ordinary capacitors and demand leveling resistors at series cells' connection and special electronic circuits for alignment of currents at parallel cells' connection. As a result, such units turn out to be rather "weighty," expensive, and not so reliable (there could be enough damage to one of the internal auxiliary elements to cause failure of the entire unit). For example, combined supercapacitor manufactured by the NessCap firm, with a capacity of 51 F and voltage of 340 V, weighs 384 kg!

At use of supercapacitor SC, the feeding circuit of the protective relays should be allocated into a separate line connected to the DC bus bar through diode D (Figure 2.44).

Due to the large capacity of the supercapacitor, the voltage reduction on feeding input of DPR at emergency mode (with loss of an external auxiliary voltage) will occur very slowly, even after passage of the bottom allowable limit of the feeding voltage.

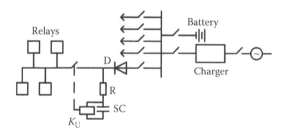

FIGURE 2.44
Example of usage of the supercapacitor as group power supply for protective relays at emergency mode with DC auxiliary voltage. K_U, voltage relay; SC, the supercapacitor.

From the personal experience of the author, cases of false operation of the microprocessor systems have been known to occur at slow feeding voltage reduction, below allowable levels. This can be explained by the existence of different electronic components of a high degree of integration serving the microprocessor, having different allowable levels of voltage feeding reduction and stopping the process of voltage reduction serially, breaking the internal logic of the DPR operation.

If such equipment is found in the DPR, used on the given substation, in parallel to the supercapacitor, it should be connected to a simple voltage monitoring relay K_U, which disconnects the supercapacitor at a voltage reduction below the lowest allowable level, for example, lower than 150–170 V.

When discussing the solutions of relay units' protection from voltage dips, it is noteworthy that there are special devices on the market that are meant to compensate voltage dips. These devices are called voltage dip compensators (VDCs). There are several principles of building VDC (Figure 2.45).

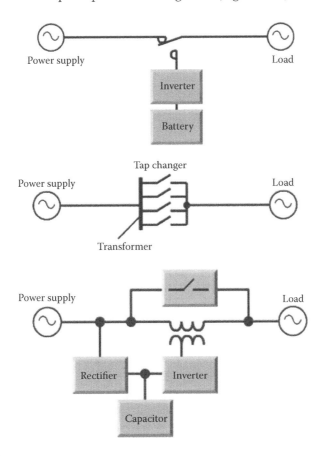

FIGURE 2.45
Several principles of building VDC.

FIGURE 2.46
Low-power VDC (up to several kilowatts).

One of them (Figure 2.45) is very similar in principle to the building of UPS units and contains a battery, an inverter, and a quick switch, which switches power to the inverter in the event of a voltage dip. This device enables compensating deep (up to zero) and lengthy voltage dips and does not require a series of accumulator batteries.

A much easier and cheaper option is a device that is part of a transformer with branches and quick semiconductor keys (Figure 2.45). This device has a limited range of voltage dip compensation.

An intermediate position in terms of quality of compensation between those aforementioned belongs to a device with a boost transformer, a reservoir capacitor, an inverter, and a quick key (Figure 2.45).

Today, there are dozens of companies in the market that manufacture many models of special VDC for any thickness of a container from small low voltage (Figure 2.46) to cabinets with a voltage rate of hundreds of kilowatts or even megawatts.

References

1. Shalin A. I. The effectiveness of the new relay protection device. *Energy and Industry of Russia*, 1 (65), 2006.
2. Prokhorov A. Intelligence is our main competitive advantage (interview with General Director of OJSC ChEAZ M.A. Shurdov). *The Equipment, Market, Supply, Price*, 4, 2003.
3. Kuznetsov M., Kungurov D., Matveev M., Tarasov V. Input circuits of relay protection devices. Issues on the high surge overvoltage protection. *News of Electrical Engineering*, 6 (42), 2006.

4. Borisov R. Negligence to the EMC may be disastrous. *News of Electrical Engineering*, 6 (12), 2001.

5. Pravosudov P. Trabtech—Technology for the surge overvoltage protection. *Components and Technologies*, 6, 2003.

6. Moore G. E. Cramming more components onto integrated circuits. *Electronics*, 38 (8), 1965, 1–4.

7. Nailen R. L. How to combat power line pollution. *Electrical Apparatus*, December, 1984.

8. Matsumoto T., Kurosawa Y., Usui M., Yamashita K., Tanaka T. Experience of numerical protective relays operating in an environment with high-frequency switching surge in Japan. *IEEE Transactions on Power Delivery*, 21 (1), 2006, 88–93.

9. Matsuda T., Kobayashi J., Itoh H., Tanigushi T., Seo K., Hatata M., Andow F. Experience with maintenance and improvement in reliability of microprocessor-based digital protection equipment for power transmission systems. Paper 34-104 on SIGRE Session, Paris, France, August 30–September 5, 1992.

10. Sowa A. W., Wiater J. Overvoltages in protective and control circuits due to switching transient in high voltage substation. Electrical Department, Bialystok Technical University, Bialystok, Poland.

11. Wiggins C. M., Thomas D. E., Nickel F. S., Wright S. E., Salas T. M. Transient electromagnetic interference in substations. *IEEE Transactions on Power Delivery*, 9 (4), 1994, 1869–1884.

12. Carsimanovic S., Bajramovic Z., Ljevak M., Veledar M., Halilhodzik N. Current switching with high voltage air disconnector. *International Conference on Power Systems Transients (IPST'05)*, Montreal, Quebec, Canada, June 19–23, 2005, Paper no. 229.

13. Rao M. M., Thomas M. J., Singh B. P. Transients induced on control cables and secondary circuit of instrument transformers in a GIS during switching operations. *IEEE Transactions on Power Delivery*, 22 (3), July 2007, 1504–1513.

14. Kuznetsov M., Kungurov D., Matveev M., Tarasov V. The problems with protection input circuits in relay protection equipment against high power impulse overvoltages. *Relay Protection and Substation Automation of Modern EHV Power Systems*, Moscow, Russia, September 10–12, 2007.

15. Stolnikov M. I. Malfunctions of relay protection at using current transformers with low magnetization curve. Report on *XVII Scientific Conference "Relay Protection and Substation Automation – 2006,"* Moscow, Russia, May 16–19, 2006.

16. Afanasiev V. V., Adoniev N. M., Dgalalis L. V. Current transformers. S.-P. *Energia*, 1980, 334pp.

17. Gurevich V. I. Microprocessor-based protective relays. How its constructed? *Electrotechnical Market*, 4–6, 2009; 1–2, 2010.

18. Benmouyal G., Zocholl S. E. The impact of high fault current and CT rating limits on overcurrent protection. SEL Publications, Schweitzer Engineering Laboratories, Inc., Quebec, Canada, 2002.

19. Zisman L., Gurevich V. Fast over current microprocessor protective relay: Theory and practice. Thesis on *International Scientific-Technical Conference "Electricity 2007,"* SEEEI, Eilat, Israel, 2007.

20. Zamora I., Mazón A. J., Valverde V., Torres E., Dyśko A. Power quality and digital protection relays. *International Conference on Renewable Energies and Power Quality (ICREPQ'04)*, Barcelona, Spain, 2004.

21. Gencer Ö. Ö., Basa Arsoy A., Özturk S., Karaarslan K. Influence of voltage harmonics on over/under voltage relay. Department of Electrical Engineering, Kocaeli University, Izmit, Turkey.
22. Gurevich V. I. Secondary power supplies: Anatomy and application. *The Electrotechnical Market*, 1 (25), 2009, 50–54.
23. Information Notice No. 94-20: Common-cause failures due to inadequate design control and dedication, Nuclear Regulatory Commission, March 17, 1994.
24. IEC 61000-4-11 Ed. 2.0 b:2004. Electromagnetic compatibility (EMC)—Part 4-11: Testing and measurement techniques—Voltage dips, short interruptions and voltage variations immunity tests.
25. IEC 61000-4-34 Ed. 1.0 b:2005. Electromagnetic compatibility (EMC)—Part 4-34: Testing and measurement techniques—Voltage dips, short interruptions and voltage variations immunity tests for equipment with input current more than 16 A per phase.
26. Iyoda I., Hirata M., Shigei N., Pounyakhet S., Ota K. Affect of voltage sags on electro-magnetic contactor, *9th International Conference "Electric Power Quality and Utilisation,"* Barcelona, Spain, October 9–11, 2007, pp. 1–6.
27. Gurevich V. *Digital Protective Relays: Problems and Solutions.* Taylor & Francis Group, London, U.K., 2011, 404pp.
28. Fox G. H., Applying microprocessor-based protective relays in switchgear with AC control power. *IEEE Transaction on Industry Applications*, 41 (6), 2005, 1436–1443.

3

Intentional Destructive
Electromagnetic Impacts

3.1 Classification and Specification of Intentional Electromagnetic Destructive Impacts

English technical literature calls intentional destructive electromagnetic impacts (IDEIs) as high-power electromagnetic threats (HPEMs). They are distinguished into two types: high-altitude electromagnetic pulse (HEMP) and intentional electromagnetic interference (IEMI).

HEMP is a very powerful electromagnetic impulse, which results from a high-altitude nuclear blast. The powerful electromagnetic impulse, resulting from a nuclear blast, was known long ago as one of the affecting factors of this blast. The fact that a nuclear blast will be accompanied by electromagnetic radiation was concluded from the scientific research of the effect of x-ray radiation performed by an American theoretical physicist Arthur Compton back in 1922 (in 1927, he was awarded the Nobel Prize for this invention). At that time, this effect didn't receive much attention and was recalled with the beginning of nuclear blast testing. The description of this is given in Ref. [1]:

> At the end of June 1946, nuclear blasts were set off near Bikini Atoll (Marshall Islands) under the name code Operation "Crossroads". During those blasts they were researching the devastating effect of nuclear weapon. The tests revealed a new physical phenomenon – development of a powerful electromagnetic emission (EME), which immediately excited the interest. The EME was especially significant with high altitude blasts. In summer of 1958 nuclear blasts were set off at high altitudes. The first series under the name code "Hardtack" was performed above Pacific Ocean near Johnston Island.
>
> During the tests two explosives of megaton class were set off, i.e. "Tack" – at 77 km altitude and "Orange" – at 43 km altitude. In 1962 high-altitude blasts continued to be set off: a blast of a 1.4 Megaton warhead was set off at 450 km altitude under the name code "Star Fish". Soviet Union also conducted a series of tests in 1961–1962. During those tests they were investigating the impact of high-altitude blasts (180-300 km)

on functionality of anti-ballistic missile equipment. During the tests powerful electromagnetic impulses were determined. They had very strong long-distance devastating effect on electronic equipment, communication and power transfer lines, radio- and radar sets on large distances.

The effect of an altitude of a 10 Mt blast on the devastating area of electronic equipment is shown in Table 3.1.

According to the classification of the International Electrotechnical Commission (IEC), there are three components of HEMP: E_1, E_2, and E_3 (Figure 3.1).

E_1 is the "quickest" and the "shortest" component of HEMP and is subject to a powerful flow of Compton's high-power electrons (the product of interaction of nuclear blast immediate radiation γ-quanta with air gas atoms), which move in the magnetic field of the earth with a velocity near the velocity of light. This interaction of quickly moving negatively charged electrons

TABLE 3.1

Area of Electromagnetic Effect of a High-Altitude Nuclear Blast

Altitude (km)	Approximate Diameter of the Area of Effect (km)
40	1424
50	1592
100	2242
200	3152
300	3836
400	4402

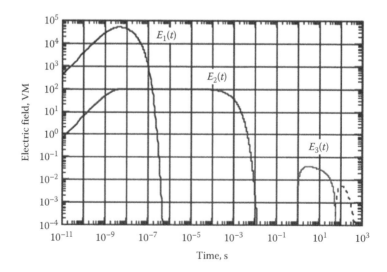

FIGURE 3.1
Parameters of components of high-altitude nuclear blast (IEC 61000-2-9).

with magnetic field produces an impulse of electromagnetic energy concentrated by magnetic field of the earth and directed from a certain altitude toward the earth.

The impulse's amplitude usually increases to its peak rating within 5 ns and reduces 50% within 200 ns. According to IEC's definition, the full duration of E_1 impulse can be about 1 μs (1000 ns). The E_1 component is subject to the most intensive electromagnetic field, which creates very high overloads in electric circuits; it creates impulse voltage up to 50 kV/m near the soil level on moderately high latitudes with the power density of 6.6 MW/m². The E_1 component causes the majority of failures of electronic equipment related to the impacts of overvoltages and the p–n junction electrical breakdown of semiconducting elements and insulation. Common arresters, which efficiently protect from atmospheric overvoltages, do not usually have enough time to come into operation and protect equipment during the E_1 component's impact, while the energy that they dissipate is not always sufficient to absorb the energy of E_1 component's impulse, which may cause destruction of common arresters.

E_2 is an "intermediate" in terms of rate of increase and duration component of HEMP, which can last according to IEC from approximately 100 μs to 1 ms. The E_2 component has a lot of similarities with electromagnetic impulses of atmospheric origin (proximal lightning). The density of field can reach 100 kV/m. Due to the similarities of E_2 component's parameters with lightning and well-developed technologies of protection against lightning, it is considered that protection against the E_2 component is not a problem. However, if E_1 and E_2 components are acting together, there will be another problem, that is, protection elements can be destroyed under the E_1 impact, while E_2 can freely enter the hardware thereafter.

The E_3 component significantly differs from the two previous components of HEMP. This is a very "slow" impulse, lasting for dozens to hundreds of seconds subject to shifting and restoration of magnetic field of the earth. The E_3 component is similar to geomagnetic storm provoked by a powerful solar flare.

Geomagnetic-induced currents are the ground currents generated by the geomagnetic disturbances in the earth's magnetosphere. These currents are also induced in long buried metal objects, such as pipelines, railroad rails, or cables. Severe disturbances in the magnetosphere of the earth arise during solar storms and the resultant emission of huge amount of ionized plasma (the so-called "solar wind") striking the earth (see Figure 3.2).

The magnetic field of the earth and its rotation around its axis generate continuous electric currents in the earth's ionosphere surrounding the planet a few hundred kilometers above the earth's surface.

The currents are maintained by the persistent generation of a large number of charged particles—ions—and free electrons radiating from the molecules of atmospheric gases broken down by solar radiation. These electric currents have a significant influence on the earth's magnetic field generation.

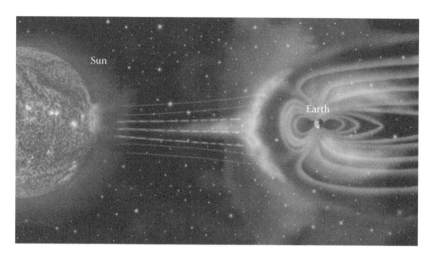

FIGURE 3.2
The distortion of the earth's magnetic field under the solar plasma ejection.

During the solar storms, extremely powerful streams of protons and electrons from the solar plasma sharply increase the electric currents flowing in the ionosphere. Aside from the rapid changes in the earth's magnetic field, such abrupt current changes generate geomagnetically induced currents and induce high currents in long power transmission lines. These induced currents are looped through the grounded neutrals of power transformers (see Figure 3.3).

Since such currents are of a very low frequency, their flow through the windings of power transformers saturates the magnetic cores of transformers and leads to the sharp decrease in transformers' impedance. As is known, the constant component of the power transformer current also appears at

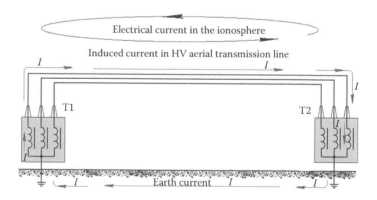

FIGURE 3.3
Diagram of currents in electric mains and ground induced by ionosphere electric currents.

the power transformer switching on, so the protective relays of power transformers are usually offset from the current constant component and do not react to it.

In addition, a constant current (or a current of a very low frequency) is practically nontransmissible through the current transformers. Thus, the traditional protective relay will not respond to the induced currents saturating the transformer, and the transformer will just burn out. In the past, there have been many cases of power transformers burnt out under the geomagnetic currents induced during solar storms. For example, in 1989, a moderate solar storm made the ultrahigh-voltage power transformers inoperative and benighted the Canadian province of Quebec for 9 h. The same storm disabled an ultrahigh-voltage power transformer at the Salem nuclear power plant in New Jersey, United States. On April 29, 1994, shortly after the beginning of the impact of the solar wind, the strong geomagnetic storm completely deactivated the ultrahigh-voltage power transformer at the Yankee nuclear power plant in Maine. On March 24, 1940, a major geomagnetic storm temporarily disrupted the electric power supply in some regions of the states of New England, New York, Pennsylvania, Minnesota, Quebec, and Ontario and incapacitated 80% of all telephone lines in Minneapolis, Minnesota [2].

The scientists of NASA expect unusually strong solar storms in 2012 (or in 2013 according to some forecasts). According to these forecasts [2], strong magnetic storms causing failures of the power systems all over the world are expected in coming years. Such failures could last from several hours to several months (due to a lack of backup power transformers in many power systems). It can result in major collapse for the humanity since we are too dependent on modern technologies and too vulnerable to disasters of this kind.

The E_3 component of high-altitude nuclear explosion [3] has a similar physical effect on power transformers and is considered by the militaries of many countries as the so-called "nonlethal weapon" aimed at destroying the infrastructure rather than killing people, which makes it particularly attractive.

The power transformers differ from the electronic devices, also exposed to damage under such impacts, in their impossibility of quick change in case of failure [3]. In the context of the foregoing, it becomes clear that it is important to protect power transformers against damage under geomagnetically induced currents of low frequency [3].

Since the 1980s, several countries have worked on creation of the so-called "Super-EME" of a nuclear explosive with improved output of electromagnetic radiation. The work is generally conducted in two directions: due to creation of a can around the explosive from a material emitting high-power γ-emission when exposing it to neutrons of a nuclear blast and due to focusing of γ-emission. Based on the calculations of the specialists, Super-EME will be able to create the field strength near the surface of the earth rated hundreds and even thousands of kilovolts per meter. It is noteworthy that military people do not conceal the information that targets of this EME weapon in future conflicts will be represented by governmental and military management systems and national infrastructure, which includes systems of power and water supply as well as communication systems.

In June of 1950, Central Physics and Technical Institute of the Ministry of Defense of the Russian Federation was established as part of the 12th Directorate of the Ministry of Defense—12th GU MO (located in central Moscow, Znamenskiy Pereulok 19, Military Unit number 31600) in Sergiyev Posad-7 town—that received the number Military Unit 51105 (today, this is Federal State Institution "12 Central Research and Scientific Institute of the Ministry of Defense of the RF") and was headed by the Member of the Academy of Science Mr. Loborev Vladimir Mikhailovich (since 2002, the head of the institute is Rear Admiral, Doctor of Technical Science, Professor Pertsev Sergey Fyodorovich). The main task of this institute was to investigate affecting factors of a nuclear blast, mainly including electromagnetic impulse as well as laser, particle beam, ultrahigh-frequency weapon, and x-ray radiation. Among the experimental instrumentations of the institute are ultrapowerful generators of impulses of high-voltage GIN-10, imitators of electromagnetic impulse of a nuclear blast IEMI-B and IEMI-BM, systems "Arterit" and "Zenith" to test equipment's resistance to electromagnetic impulses and nuclear reactor "BARS," etc. The 12th GU MO up to now is probably the most secretive organization of the Soviet/Russian Armed Forces, even more than the GRU (Main Intelligence Directorate of the General Staff of the Armed Forces of the Russian Federation) or the Strategic Rocket Forces.

A powerful EME can be created not only as a result of a nuclear blast. Modern advancements in the area of nonnuclear generators of EME allow to produce them rather compact and to use them with common and highly accurate delivery means. That's why the issues of protection from EME's impact will be of interest for specialists regardless of the outcome of negotiations about nuclear weapon.

IEMI is the second type of IDEIs, which has nothing to do with nuclear blast.

The first technical ideas about a possibility to create nonnuclear explosively pumped flux compression generator (EPFCG) were articulated by the Russian Academic Andrei Sakharov in 1951, when he was working on a nuclear warhead in Arzamas-16 (now Russian National Research and Scientific Institute of Experimental Physics [RNRSIEP]) [4].

The first experimental work to obtain an ultrapowerful impulse magnetic field by their compression was started at this institute by Robert Lobarev who managed to obtain impulse magnetic fields rated 1.5 million Gauss (compared to magnetic fields of the earth of 0.3 G at the equator and 0.7 G in polar regions) in 1952. Later, this work was continued by Alexander Pavlovsky and Vladimir Chernyshov from the same institute. The team guided by A. Pavlovsky managed to build an explosive generator (Figure 3.4) with

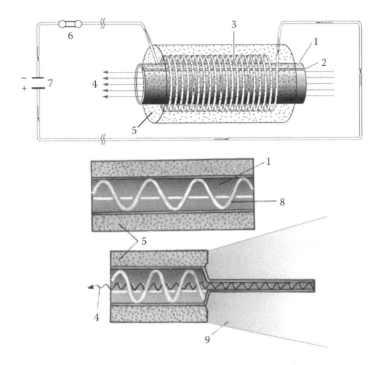

FIGURE 3.4
EPFCG. (1) Electromagnetic resonator. (2) Section. (3) Coil streamlined by current. (4) Directed electromagnetic radiation. (5) Explosive. (6) Switch. (7) Energy storage device (capacitor). (8) Standing wave. (9) Flying explosion debris.

impulse current rated 200 million Amperes, which could generate magnetic field of 10 million Gauss.

The EPFCG was represented by a ring of explosive material surrounding a copper coil. A set of simultaneously setoff primers initiated detonation directed to the axis. At a certain moment synchronized with a set off a powerful capacitors is discharged and its current creates a magnetic field inside the coil. The shock-wave crushes and "short-circuits" the coil's windings by huge pressure and transforms the coil into a tube, closing its field inside it. The current circuit is compressing at high speed—several kilometers per second—depending on the type of the explosive material. As we know from physics classes, the intensity of magnetic flux created by the circuit in this case is proportional to the speed of inductance changing in time. Since the coil's size is changing with big speed during collapse of the circuit, the amplitude of magnetic field is also huge (dozens of millions of amperes). At this moment, one of the ends of the resonator cavity is destructed by means of an explosive cartridge, and the shock wave converging to a point will be reflected and rush backward, changing the field abruptly. Meanwhile, the coincident wave will be converted into a progressive wave developing a huge impulse power, which will result in generation of impulse flux of radio-frequency electromagnetic radiation. The field will be changing within parts of nanoseconds; however, it will not follow the sinusoidal law with a period equal to compression/rarefaction, but abruptly, and this means that the function describing its changes will include a lot of frequencies. That's why the shock-wave source is highly broadband and radiates in the range from hundreds of megahertz to hundreds of gigahertz, while the impulse is lasting hundreds of microseconds.

Almost simultaneously but independently from them, a team managed by Max Fowler from Los Alamos National Laboratory obtained same fields. Both scientists met in Novosibirsk in 1982 at *the International Conference on High-Power Magnetic Fields*. In 1989, a group of Soviet scientists managed by A. Pavlovsky was welcomed in Los Alamos National Laboratory.

Since the very beginning of this work, both scientists of the United States and USSR and politicians understood that this type of sources of high-power electromagnetic impulses can create grounds to develop a new type of weapon. This is evidenced by the speeches of N. S. Khrushchev in the 1960s where he mentioned some sort of "fantastic equipment" developed by Soviet scientists.

The EPFCG as a separate armor unit to create ultrapowerful electromagnetic impulses was officially introduced by the head of the laboratory of special-purpose ammunition of the Central Research and Scientific Institute of Chemistry and Mechanics Doctor of Technical Science Mr. A. B. Prishchepenko after successful tests at the training range of Krasnoarmeysk Research and Scientific Institute, Geodeziya, Moscow region (now known as Federal State-Founded Enterprise Research and Scientific Institute, Geodeziya—the main firing field of Russian military industries; last director, PhD Vagin Alexander Vasilyevich) on March 2, 1983. Later on, the Corresponding Member of the Academy of Military Science A. B. Prishchepenko articulated general principles of tactical deployment of electromagnetic ammunition [5,6].

Today, intensive research in the field of IEMI is conducted in several directions, and nonnuclear EPFCGs are not the only type of nonnuclear electromagnetic weapon.

There is a wide range of high-power microwave (HPM) equipment: relativistic klystron tubes and magnetrons, reflex triodes, backward-wave tubes (BWTs), gyrotrons, virtual cathode oscillators (vircators), and others (Figure 3.5).

Vircators can produce very powerful individual energy impulses, which are structurally simple, not large in terms of their size, durable, and capable of working in a relatively wide band of microwave range frequencies. The fundamental idea behind a vircator is to accelerate a powerful flow of electrons by a grid anode. This powerful flow of electrons initially leaves the cathode (metal cylinder-shaped rod, several centimeters in diameter [Figure 3.5]) as affected by high-voltage impulse (hundreds of kilovolts), which provides the emission of electrons' explosive behavior. Significant number of

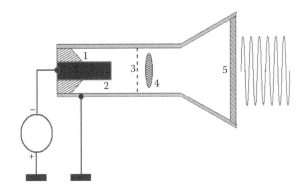

Inductive power storage VIRCATOR		Capacitive power storage VIRCATOR	
Voltage pulse (300 ns), kV	400	Voltage pulse (100 ns), kV	600
Current, kA	12	Current, kA	18
Power irradiation, MW	350	Power irradiation, MW	500
Duration of irradiated pulse, ns	200	Duration of irradiated pulse, ns	80
Output frequency, GHz	3.1	Output frequency, GHz	3.1

FIGURE 3.5
Powerful vircators. (1) Insulator. (2) Metal cathode. (3) Grid anode. (4) Virtual cathode. (5) Dielectric window.

electrons goes through the grid anode creating a charge cloud behind the anode. Under specific conditions, this charge cloud will be oscillating in the area of anode (Figure 3.6).

The microwave field obtained at the oscillation frequency of electronic cloud is emitted into space through a dielectric window. Initial current rating in vircators at which generation starts amounts to 1–10 kA. Vircators (Figure 3.5) are the most suitable for generation of nanosecond impulses in the longwave part of S-band. They produced power from 170 kW to 40 GW in S-band and L-band under laboratory conditions. Based on the published data, an experimental apparatus, which can develop power of about 1 GW (265 kV, 3.5 kA), can affect electronic equipment from 800 to 1000 m.

FIGURE 3.6
Relativistic microwave high-power generators as part of gyrotrons, vircators, and BWTs, developed at different research and scientific institutes of Russia.

Even well-known high-voltage impulse Marx's generators (Figure 3.7), which contain a set of high-voltage capacitors and dischargers (80 similar units), can be used as powerful sources of microwave radiation. All the condensers in this device are charged concurrently from a high-voltage source, whereas during a disruptive discharge of controlled dischargers, all these capacitors appear to be connected in series.

FEBETRON-2020 (the mobile generator) (Figure 3.7) generates current impulses of 6 kA at 2.3 MV. As a result, powerful electromagnetic impulses are emitted. Based on the Marx's design, Applied Physical Electronics (United States) developed a series of powerful very compact generators for voltages up to 1 MV with emitted peak voltage up to 6 GW (Figure 3.8). Equipped by a dish-shaped antenna (Figure 3.9), these devices can emit devastating effect for electronic equipment directed microwave energy of extremely high power.

Another developmental field of IEMI is the so-called beam weapon. This weapon is using straight beams of charged or neutral particles, generated by means of different types of accelerating units, both land- and space-based. The work on creation of beam weapon gained the widest spread after Ronald Reagan (former U.S. president) announced the Strategic Defense Initiative Program in 1983. The center of scientific research in this area became the Los Alamos National Laboratory and the Livermore National Laboratory,

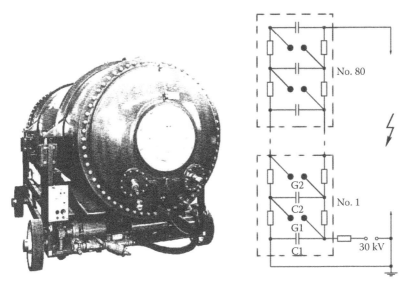

FIGURE 3.7
American FEBETRON-2020 as part of Marx's generator and its simplified layout.

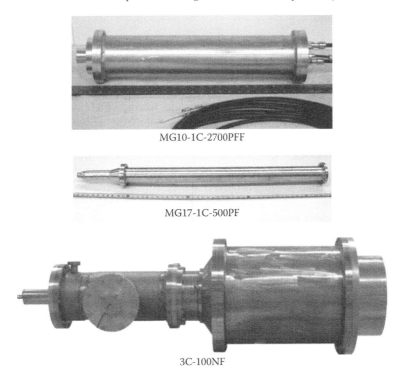

FIGURE 3.8
Compact high-power Marx's generators: 300 kV, 1 GW (MG10-1C-2700PFF); 510 kV, 400 MW (MG17-1C-500PF); 600 kV, 6 GW (MG30-3C-100NF).

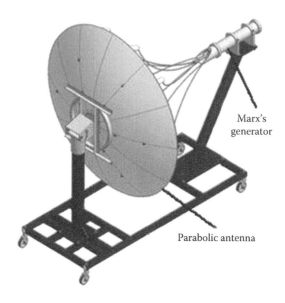

FIGURE 3.9
A powerful generator of directed microwave radiation based on compact Marx's generator with parabolic antenna.

while in Russia, it was the aforementioned Federal State Institution 12 Central Research and Scientific Institute of the Ministry of Defense of the RF. According to several scientists, they made successful attempts there and obtained a stream of high-powered electrons, which outcompetes the power of the stream obtained in research accelerating units in several hundred times.

The same laboratory experimentally determined in the framework of "Antigone" program that the electronic beam almost perfectly (without diffusion) dissipates across the ionized channel initially created by the laser ray in the atmosphere.

Powerful compact sources of radiation are very dangerous as they can be mounted in a covered truck's trailer and even in a minivan.

For example, Tomsk Research and Scientific Institute of High-Current Electronics, established in 1977 with a purpose to research the methods of generation of high-powered (gigawatt and terawatt) electric impulses under the guidance of Academic G. A. Mesyats, developed rather powerful compact generators (100–1000 MW) of linearly polarized single-directional beams of ultrabroadband electromagnetic radiation with a nanosecond and subnanosecond duration of impulse to affect electronic equipment (Figure 3.10). Moreover, these devices can be freely purchased at the institute at a rather decent price of USD 40,000–60,000 and installed in a minivan or a small trailer. Contact information to make the orders is provided on the official website of the institute.

FIGURE 3.10
Compact powerful ultrabroadband radiation sources with a power up to 1 GW developed by Tomsk Research and Scientific Institute of High-Current Electronics SB RAS.

Similar mobile and handheld sources are developed and produced in the United States (Figure 3.11).

According to the mass media information, Boeing conducted trials at the training range in Utah (Utah Test and Training Range [UTTR]) under combat conditions in October 2012. They were testing a missile developed

FIGURE 3.11
Compact source of powerful directed ultrashortwave (95 GHz) radiation developed by American Sandia National Laboratories using technologies of Raytheon Company (mentioned previously) and powerful sources of directed radiation mounted on chassis of Hummer vehicle and armored vehicle Stryker. Even more powerful unit will be mounted on board a plane AC-130.

in the framework of Counter-Electronics High-Power Microwave Advanced Missile Project (CHAMP) program (the design of an improved missile with a powerful microwave radiating unit to fight with radio-electronic hardware). During the trials, the missile CHAMP, which was flying according to a pre-scribed program, generated powerful energetic impulses putting electronic subsystems out of action and destroying data without any physical damage. The CHAMP missile demonstrated the ability to perform selective and pre-cise attacks at several targets during one flight using a powerful microwave radiation unit HPM. Flight control was performed from the Hill air base.

CHAMP should become a nonlethal alternative to arms systems of kinetic action and traditional means of devastation of explosive action when strik-ing out at enemy's facilities equipped with radio-electronic equipment. The missile allows to set facilities out of action by destroying its electronic equipment and causing minimum incidental physical damage to the area of dislocation.

According to the program manager at Boeing Phantom Works, Mr. Kate Coleman, this technology will pioneer the new era of modern war. In near future, it can be used to set electronic equipment and information systems of enemy out of order without any aviation- or land-based forces. According to Boeing representatives, the project is an adaptation of technology of directed energy developed by the laboratory of the U.S. Air Force for a plat-form created by the missile's company, and it will be the foundation for the creation of a new family of high-efficient systems of armor of nonlethal action.

Being the general subcontractor, Boeing manufactures an air platform and performs final integration of all systems. Raytheon Company provides a source of HPM radiation, whereas the National Laboratory Sandia provides powering systems under a separate contract with the National Research and Scientific Laboratory of the U.S. Air Force.

The TV ad of Boeing shows a cruise missile (Figure 3.12) flying above a city and "switching" its lights off. In particular, you can see a main control board of a power plant and all its lights go off, while the missile is flying.

Due to high complexity of the issue, electromagnetic weapon is devel-oped by a short list of companies. The world leaders include American companies Northrop Grumman, Lockheed Martin, Raytheon, and ITT and a British Company BAE Systems. In Russia, the leading manufacturer and developer of devices for radio-electronic warfare is the public cor-poration, Corporate Group Radio-Electronic Technologies. The Holding was established in 2009 and includes 18 enterprises, such as research and scientific institutes, design offices, and production facilities, which specialize in the creation of air-, sea-, and land-based means for radio-electronic warfare. In the framework of the state arms program for 2011–2020 (SAP-2020), The Holding increases its share in the market of means for radio-electronic warfare.

FIGURE 3.12
Advertisement of a cruise missile with an electromagnetic battle part CHAMP by Boeing.

FIGURE 3.13
Mobile station of radio-electronic suppression 1L269 "Krasukha-2."

Recently, successful state trials have been conducted and industrial pro-duction of "Moscow-1," "Krasukha-2," "Krasukha-4," "Rtut" ("Mercury"), and other complexes has been launched (Figure 3.13). All the aforementioned complexes were developed by the Russian National Research and Scientific Institute, Gradient, which is part of The Holding.

FIGURE 3.14
Mobile system of highly powerful long-distance microwave radiation "Ranets-E."

Mobile microwave system "Ranets-E" (developed by Moscow Radiotechnical Institute) with impulse radiating power of 500 MW ensures guaranteed destruction of electronics from 12 to 14 km, while serious interference in the operation of electronic equipment can be observed from distances up to 40 km (Figure 3.14).

In fact, this is a powerful microwave especially intended to destroy electronic equipment.

In the 1980s a renowned Soviet scientist professor I. V. Grekhov from Ioffe Physical-Technical Institute of the Russian Academy of Sciences (RAS) (St. Petersburg) conducted theoretical and experimental works on the formation of high voltage lasting for nanoseconds of voltage swings on usual high-voltage semiconducting diodes [7–9]. The occurrence of overloading during switching of power semiconducting diodes from conducting direction to blocking direction is well known in engineering. Engineers try to reduce this event by many different ways, since these overvoltages reduce reliability of diodes and other elements of electronic circuits. In the work guided by I. V. Grekhov, scientists tried to increase this effect and use it to generate powerful impulses lasting for nanoseconds by using semiconducting high-voltage diodes as current breakers in powerful impulse systems with inductive energy storing device.

Later on, this research was continued at the Ural Branch of the Institute of Electrophysics of the RAS (Yekaterinburg). Experiments conducted in 1991–1992 by S. K. Lyubutin, S. N. Rukin, and S. P. Timoshenkov on usual high-voltage rectifying semiconducting diodes showed that under specific combination of density of conducting and blocking current and time of its flowing through a semiconducting structure of the diode, the time of decline of blocking current reduces to dozens and less nanoseconds. The rating of current density amounts to dozens of kiloamperes per square centimeter, while the time of current flowing lies in a range of hundreds of nanoseconds. This new effect of nanosecond drop of very dense currents in semiconductors was later called silicon on sapphire (SOS) effect (from semiconductor opening switch) [10].

Later on, a special semiconducting structure with ultrahard recovery mode was developed. This system allowed creating high-voltage semiconducting breakers of a new class, that is, SOS diodes having operational voltage of hundreds of kilovolts, communication current of dozens of kiloamperes, and switching time of several nanoseconds as well as switching frequency—kilohertz.

A typical design of a SOS diode is a consecutive assembly of elementary diodes fixed by dielectric pins between two plate electrodes (Figure 3.15) [11].

On the basis of SOS diodes, the Ural Branch of the Institute of Electrophysics of the RAS developed a series of reusable compact SOS generators of electromagnetic impulses in the nanosecond range with outstanding parameters for semiconducting switches (Figure 3.16). According to information published on the website of the institute, these generators are based on using of solid-state system of switching of energy, where thyristors and transistors are used on the input unit, magnetic switches on the intermediate unit of energy compression, and semiconducting current interrupter on SOS diodes in a back-end amplifier. These generators have the following range of output parameters: range of voltage impulse

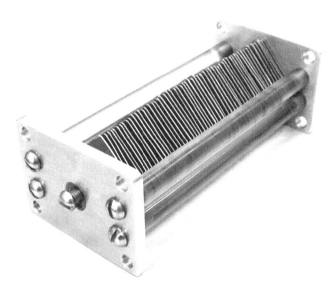

FIGURE 3.15
SOS diode type SOS-180-12 (180 kV, 12 kA).

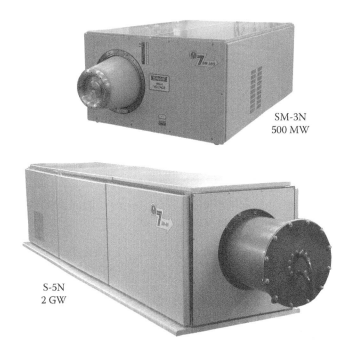

SM-3N
500 MW

S-5N
2 GW

FIGURE 3.16
Several types of SOS generators, developed by UO of the Institute of Electrophysics of the RAS.

50 kV to 1 MV, impulse current 1–10 kA, peak power 100 MW to 4 GW, duration of impulse 3–60 ns, and frequency of impulses hundreds of Hz to several kHz.

Some countries develop compact electromagnetic rifles of relatively low power. However, they will be able to set electronic equipment out of order from 100 m distance. Not only military but also police take interest in these. A modern car full of different electronic devices can also be a target like any other sophisticated system. American company Eureka Aerospace developed and launched production of electromagnetic "stopper" of a driving car (electromagnetic pulse [EMP] car stopper). The action of this weapon is based on damaging microprocessor, ignition system, fuel injection, and other electronic systems of a modern car. What will happen when terrorists will possess this weapon (and this will happen sooner or later) and they will use it against electric power facilities? However, they do not need to put a lot of efforts into searching this weapon. Many popular technical magazines provide description on how to assemble these devices at home (Figure 3.17).

These rather powerful homemade sources of directed ultrahigh-frequency radiation installed in a covered trailer of even a van (Figure 3.18) can be extremely dangerous for electric power facilities (and not only for them).

FIGURE 3.17
Homemade directed microwave generators. The description of these is given in popular technical magazines.

You can easily imagine how this vehicle with a turned-on generator is passing by a modern substation with lots of working digital protection and control devices, but it is difficult to imagine what will happen in the power system. In fact, one of these vehicles can "switch off" several substations. And now imagine if we deal with several vehicles.

Moreover, the industry also chips in into "common efforts" producing such devices, assembled in a small suitcase (Figure 3.19) as if they were specially designed for terrorists. It is not in vain that one of the congressmen of the United States mentioned it in its report.

I think it is a perfect situation to quote Winston Churchill who said many years ago: "The Stone Age can come back on shiny wings of science."

FIGURE 3.18
A source of directed ultrahigh-frequency radiation installed in a pickup with a plastic box.

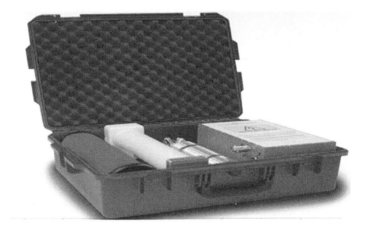

FIGURE 3.19
Marching "destructor of electronics" called "2100 Series Suitcase" based on Marx's generator, which is manufactured by the Applied Physical Electronics Company.

3.2 IDEI's Impact on Digital Protection Relays

Electromagnetic radiation can enter electronic equipment in several different ways, such as through different antenna devices and cable inputs, power supply systems, as well as current induced in wrapping as well as radiation entering through doors and windows produced from nonconductive materials and ventilation ducts. Current induced by EME in land and buried power supply cables of hundreds and thousands of kilometers in length can achieve thousands of amperes, while voltage in opened circuits of these cables can reach as high as thousands of volts. As for antenna inputs, the length of which does not exceed several dozens of meters, and the rating of currents induced by EME may be hundreds of amperes. The EME entering directly through the elements of constructions made of dielectric materials (nonshielded walls, windows, doors, etc.) can induce currents in internal cabling up to dozens of amperes. Long aerial power transfer lines pose the biggest danger as they absorb emission from large areas and deliver it to the destination point, that is, to inputs of very sensitive electronic equipment. Availability of transformers on this way (measuring and supply) has almost no effect on this process due to significant internal capacitance between the primary and secondary windings. Since low-current circuits and radio-electronic appliances are working at voltage ratings of several volt and current ratings of several dozens of milliamperes, it is necessary to reduce current and voltage ratings in several times at their inputs to ensure their reliable protection from EME. Besides digital protective relay (DPR), optical data transfer systems, which are widely used in relay protection (RP), are also rather sensitive to EME. In fact, I refer to controllers, which transform electric signals into optical at one end of fiber-optic communication system (FOCS) and restore it from optical at the other end of FOCS. For example, testing of fiber optic communications for utility systems (FOCUS) multiplexer [12] for its correspondence to IEC standards in terms of electromagnetic compatibility showed that is does not always resist standard impacts without failures and damages. The supervisory control and data acquisition (SCADA) system with its large number of digital sensors and measuring converters united in a common computer network is another example susceptible even to weakened EME.

If a possibility of use of a high-altitude nuclear blast for electromagnetic destruction of national power system can be hypothetical, the attack of terrorists to local power system by means of simultaneous impact on several most important units by means of nonnuclear EME sources is very likely at any moment.

Systems of data transfer using protocols with broadband frequencies are the most susceptible to intentional electromagnetic impacts. These systems include ATM 155, Fast Ethernet, and Gigabit Ethernet. The latter is explained by insignificant power difference of a friendly signal and distortions in the upper part of the spectrum. Transition from coaxial cable to simple twisted-pair wire in order to reduce cable costs (a tendency all over the places) results in even higher susceptibility of the system. It should be noted that the Ethernet powered by twisted-pair cable is now used in RP and according to smart grid strategy, its use to control electric power facilities will be broadening further.

Discrete electronic elements are much more resistant to overvoltages and other detrimental impacts rather than integral microcircuits [13]. According to Ref. [14], 75% of all damages of microprocessor-based devices are due to the impact of overvoltages. These overvoltages with amplitudes from dozens to several kilovolts, which appear as a result of switching processes in circuits or under the impact of electrostatic discharge, are "deathly" for internal microelements of microcircuits and processors. According to Ref. [14], conventional transistors (discrete elements) can resist 70 times higher voltage of electrostatic discharge than, for instance, a memory microchip (erasable programmable read only memory [EPROM]) of a microprocessor system. Computerized industrial equipment (including DPR) is especially susceptible to EME impact since it is based on metal oxide semiconductor (MOS) structures of high density, which are very sensitive to impacts of high-voltage transient processes. MOS structures feature very low level of energy (voltages about dozens of volts) necessary for their damage of full demolishing.

There are three stages of degradation of semiconducting instruments under the impact of powerful EME: failure in operation, sustainable change of parameters, and irreversible catastrophic faults. Irreversible fault of semiconductors is mainly due to their overheating or field rupture [15–17]. The damage of a microprocessor or memory elements induced by weakened electromagnetic impacts may be of concealed nature [18]. Such damages are not revealed by any test and may come to a surface very unexpectedly. Besides, accidental, reversible failures subject to self-induced changes in memory cells' content called "soft errors" may happen as a result of EME weakened by reduced with protective measures impacts. This type of mistakes (reversible, self-restoring failures of operation) was not known until recently for electronic instruments produced on the basis of discrete semiconducting elements or ordinary microcircuits.

The recent progress in the area of nanotechnologies resulted in significant shrinking of size of semiconducting elements (i.e., several or even parts of a micron), reduction of thickness of layers of semiconducting and insulation materials, reduction of operating voltage, reduction of electric capacity of individual memory cells, and increase of density of elementary logical cells in one device. Altogether, this led to sharp increase of sensitivity of memory elements to electromagnetic impacts. The problem is worsened by the fact that there is a tendency to widen the use of memory elements in modern microprocessor-based structures. The majority of modern integral microcircuits of high level of integration that are part of a microprocessor-based device contain built-in memory elements of a rather big capacity and the failure of which is not controlled at all. The problem of sharp increase of sensitivity to electromagnetic impacts is actual not only for memory elements but also for high-speed logical elements, comparators, etc., in other words for total modern microelectronics.

The level of damage depends on the insensitivity of each circuit component and the energy of the powerful interference as a whole, which can be absorbed into the circuit without the appearance of any defect or failure. For example, although the switching noise caused by the inductive load with an amplitude of 500 V is a twofold voltage surge, it is unlikely to lead to the failure of an electromagnetic relay with a 230 V AC coil due to its insensitivity to this kind of interference and its short duration (it lasts only several microseconds). The situation is different if the chip is powered from a 5 V

DC source. The impulse interference with an amplitude of 500 V is 100-fold higher than the supply voltage of the electronic component and leads to the inevitable failure and the subsequent destruction of the device. Surge resistance of the chips is several orders of magnitude lower than that of the electromagnetic relays [19]. Long-term statistics confirms that the number of such damages doubles every 3–4 years [19]. This statistic is in good agreement with the so-called Moore's law [20,21] that in 1965 showed that the number of semiconductor components in microchips doubles roughly every 2 years and this trend has remained valid for many years. If some 10 years ago the so-called transistor–transistor logic (TTL) chip contained 10–20 elements per square millimeter and had a typical supply voltage of 5 V, now the popular chip can contain nearly a hundred of complementary metal–oxide–semiconductor (CMOS) transistors on every square millimeter of the surface and has a supply voltage of only 1.2 V. The up-to-date solid-state technologies raise the number up to tens of thousands of elements per cubical millimeter. It is obvious that such chips would require even lower supply voltage. Today, already only 0.9 V is needed for power supply of operational amplifier LMV951, and it is even more obvious that such improved microelectronics integrity reduces insensitivity of its components to high voltage surge due to the reduced distance between electroconductive elements, lower thickness of insulating layers, and reduced operating voltage of semiconductor elements.

Recent trends of technological evolution and ever-growing electromagnetic vulnerability of national infrastructures (power and water supply, communications, etc.) have come under military consideration long since. Military research centers of almost all developed nations have carried out intensive research and development on special weapons capable of destroying electronic equipment. The mass media have published dozens of articles discussing methods for increasing efficiency of electromagnetic actions aimed at destroying electronic equipment [22–34].

The ability of Faraday's cage to protect from electromagnetic impacts is well known. Concrete buildings contain a grounded cage, protection relays are located in metal cabinets, and DPRs have metal casing. One would think it is not a cage, but a "Faraday's nesting doll matryoshka." However, everything is not that easy. First of all, high-frequency impulses are easily getting through the holes in Faraday's cage, through any nonmetal inserts and windows, and through glass windows of buildings and ventilation systems. With this partially weakened impact of EME to semiconducting instruments, there were cases of partial damage of their p–n-junctions, which resulted in changes of their specifications and emergence of "blinking" failures in the instruments' operation. These failures connect significant number of resources intended for maintenance. Moreover, they limit the surety in reliability of equipment. These "blinking" failures are sometimes difficult to reveal, and this makes it necessary to withdraw equipment out of service with long idle time for diagnostics of damages.

This factor should also be considered when evaluating protection of equipment from electromagnetic attack since partial or insufficient protection can cause additional problems.

The second problem is known as "delayed action of EME" and represents a very dangerous feature of IDEI. This effect appears within the first several minutes after detonation of a nuclear blast or the blast of an electromagnetic bomb. At this time, EME entering though electric systems creates local electromagnetic fields. During field falloff, there will be acute drops of voltage, which dissipate in a form of waves along the power supply wires for rather long distances from a point of initial EME. Thirdly, outside cables and wires, which come out of RP cabinets and buildings and run for multiple kilometers, deprive both buildings and RP cabinets of even weakening effect.

3.3 Main Regulatory Documents in the Field of IDEI

Due to comprehension of seriousness of the IDEI problem, this topic is continuously discussed by such organizations as the International Electrotechnical Commission (IEC), International Council on Large Electric System (CIGRE), and special commission at the U.S. Congress and European bodies. Obviously, in order to achieve productivity in this area, there should be corresponding standards and other regulatory and technical documentation. Some of these standards have already been developed by IEC:

IEC TR 61000-1-3 Electromagnetic compatibility (EMC)—Part 1–3: General—The effects of high-altitude EMP (HEMP) on civil equipment and systems.

IEC 61000-1-5 High power electromagnetic (HPEM) effects on civil systems.

IEC 61000-2-9 Electromagnetic compatibility (EMC)—Part 2: Environment—Section 9: Description of HEMP environment—Radiated disturbance. Basic EMC publication.

IEC 61000-2-10 Electromagnetic compatibility (EMC)—Part 2–10: Environment—Description of HEMP environment—Conducted disturbance.

IEC 61000-2-11 Electromagnetic compatibility (EMC)—Part 2–11: Environment—Classification of HEMP environments.

IEC 61000-2-13 Electromagnetic compatibility (EMC)—Part 2–13: Environment—High-power electromagnetic (HPEM) environments—Radiated and conducted.

IEC 61000-4-23 Electromagnetic compatibility (EMC)—Part 4–23: Testing and measurement techniques—Test methods for protective devices for HEMP and other radiated disturbances.

IEC 61000-4-24 Electromagnetic compatibility (EMC)—Part 4: Testing and measurement techniques—Section 24: Test methods for protective devices for HEMP conducted disturbance - Basic EMC Publication.

IEC 61000-4-25 Electromagnetic compatibility (EMC)—Part 4–25: Testing and measurement techniques—HEMP immunity test methods for equipment and systems.

IEC 61000-4-32 Electromagnetic compatibility (EMC)—Part 4–32: Testing and measurement techniques—High-altitude electromagnetic pulse (HEMP) simulator compendium.

IEC 61000-4-33 Electromagnetic compatibility (EMC)—Part 4–33: Testing and measurement techniques—Measurement methods for high-power transient parameters.

IEC 61000-4-35 Electromagnetic compatibility (EMC)—Part 4–35: Testing and measurement techniques—HPEM simulator compendium.

IEC 61000-4-36 Electromagnetic compatibility (EMC)—Testing and measurement techniques—IEMI Immunity Test Methods for Equipment and Systems.

IEC/TR 61000-5-3 Electromagnetic compatibility (EMC)—Part 5–3: Installation and mitigation guidelines—HEMP protection concepts.

IEC/TS 61000-5-4 Electromagnetic compatibility (EMC)—Part 5: Installation and mitigation guidelines—Section 4: Immunity to HEMP—Specifications for protective devices against HEMP radiated disturbance. Basic EMC Publication.

IEC 61000-5-5 Electromagnetic compatibility (EMC)—Part 5: Installation and mitigation guidelines—Section 5: Specification of protective devices for HEMP conducted disturbance. Basic EMC Publication.

IEC 61000-5-6 Electromagnetic compatibility (EMC)—Part 5–6: Installation and mitigation guidelines—Mitigation of external EM influences.

IEC 61000-5-7 Electromagnetic compatibility (EMC)—Part 5–7: Installation and mitigation guidelines—Degrees of protection provided by enclosures against electromagnetic disturbances (EM code).

IEC 61000-5-8 Electromagnetic compatibility (EMC)—Part 5–8: Installation and mitigation guidelines - HEMP protection methods for the distributed infrastructure.

IEC 61000-5-9 Electromagnetic compatibility (EMC)—Part 5–9: Installation and mitigation guidelines—System-level susceptibility assessments for HEMP and HPEM.

IEC 61000-6-6 Electromagnetic compatibility (EMC)—Part 6–6: Generic standards—HEMP immunity for indoor equipment.

Some of the standards are still under development stage.

American Association of Electrical Engineers offers its own standard IEEE: IEEE P1642 Recommended Practice for Protecting Public Accessible Computer Systems from Intentional EMI.

A similar document has been developed by the European Commission: Topic SEC-2011.2.2-2 Protection of Critical Infrastructure (structures, platforms and networks) against Electromagnetic (High Power Microwave [HPM]) Attacks.

There is a separate working group in SIGRE dealing with the following topic: CIGRE WG C4.206 Protection of the High Voltage Power Network Control Electronics against IEMI.

Russia has also put some efforts into common goal offering its standard: State Standard R 52863-2007 Protection of information. Automated systems in guarded construction trials to resist intentional powered electromagnetic impacts. General requirements.

It should be admitted that this Russian standard has a very narrow area of application, but its existence indicates at a certain concern of specific bodies about IDEI problem.

It should be mentioned that in spite of so big amount of regulatory documents for the most part, they provide rather vague and not specific recommendations, which, of course, can be admitted considering the specific character of the topic.

We can only hope that "too many cooks" will not spoil the "broth" and the efforts of these international organizations will be successful and will result in clear and specific standards, based on which it will be possible to develop requirements to parameters of instruments and methods of their testing for resistance to IDEI.

In conclusion, I'd like to note an oracular utterance of Winston Churchill who said many years ago that *the latest refinements of science are linked with the cruelties of the Stone Age and to be amazed at his prophecy.*

References

1. Belous V. Threats of EMP weapons used for terrorist activities and military operations. *Control of Nuclear Weapon*, 1 (75), 2005, 133–140.
2. Buralkov A. A., Kibardin V. V. About effects of solar storms on reliability of energy systems. *Scientific and Technical Congress "Worldwide Power Industry,"* Krasnoyarsk, Russia, June 2010, pp. 32–33.

3. Gurevich V. I. Protection of power transformers against geomagnetically induced currents. *Serbian Journal of Electrical Engineering*, 8 (2), 2011, 333–339.

4. Sakharov A. D. Magnetic explosion generators. *The Advances in Physical Sciences*, 83–84 (4), 1966.

5. Prishchepenko A. B., Kiseljov V. V., Kudimov I. S. Radio frequency weapon at the future battlefield: Electromagnetic environment and consequences. *Proceedings of the EUROEM94*, Bordeaux, France, May 30–June 3, 1994, Part 1, pp. 266–271.

6. Prishchepenko A. B. *The Explosions and Waves. Explosion Sources of Electromagnetic Radiation of Radio Frequency Range.* Course Book for 170103 Department "Means of Defeat and Ammunition." BKL Publishers, Moscow, Russia, 2008. ISBN 978-5-94774-726-3.

7. Grekhov I. V., Efanov V. M., Kardo-Sysoyev A. F., Shenderey S. V. Establishment of high-voltage nanosecond voltage drops on semiconducting diodes with a drift restoration mechanism. *Letters to TFJ*, 9 (7), 1983, 435–439.

8. Tuchkevich V. M., Grehov I. V. *The New Principles of High Power Switching by Semiconductor Devices.* Saint-Petersburg Science, 1988, 117pp.

9. Grekhov I. V. Impulse switching of the high powers by semiconductor devices. In *Physics and Engineering of Powerful Pulse Systems*, Velihov E. P. (ed.), Energoatomizdat, Moscow, Russia, 1987, p. 237.

10. Slusar V. I. The generators of super powerful electromagnetic impulses in the information wars. *Electronics: The Science, Engineering, Business*, 5, 2002, 60–67.

11. Slovikovsky B. G. Compact generators of high voltage nanosecond impulses on the basis of SOS-diodes. Synopsis of the PhD dissertation, Ural Branch of the Institute of Electrophysics of the Russian Academia of Science, Yekaterinburg, Russia, 2004.

12. Gurevich V. I. Optoelectronic transformers: The panacea or the partial solution for the partial problems. *News in Power Engineering*, (2), 2010, 24–28.

13. Electromechanical vs. solid state relay characteristics comparison. Application Note 13c3235, Tyco Electronics, Berwyn, PA.

14. Gurevich V. *Electronic Devices on Discrete Components for Industrial and Power Engineering.* CRC Press, Boca Raton, FL, 2008, 420pp.

15. Bludov S. B., Gadetskiy N. P., Kravtsov K. A. et al. Generating powerful ultra-short microwave pulses and impact on electronics, *Plasma Physics*, 20 (7, 8), 1994, 712–717.

16. Panov V. V., Sarkisyan A. P. Some aspects of a problem of creation of microwave means for functional destruction. *Foreign Radioelectronics*, 10–12, 1993, 3–10.

17. Antipin V. V., Godovitsyn V. A., Gromov D. V., Kozhevnikov A. S., Ravayev A. A. Impact of powerful pulse microwave interferences on semiconducting devices and chips. *Foreign Radioelectronics*, 1, 1995, 37–53.

18. Phadke A. G. Hidden failures in electric power systems. *International Journal of Critical Infrastructures*, 1 (1), 2004, 64–75.

19. Pravosudov P. Trabtech—Technology for protecting electrical equipment against pulse surge. *Components and Technologies*, 6, 2003.

20. Moore G. E. Cramming more components onto integrated circuits. *Electronics*, 38 (8), 1965.

21. Nailen R. L. How to combat power line pollution. *Electrical Apparatus*, December 1984.

22. Bludov S. B. et al. Generating high-power ultra-short microwave pulses and theirs impact on electronics. *Plasma Physics*, 20 (7, 8), 1994, 712–717.

23. Didenko A. N., Sulashkin A. S., Fortov V. E., Yushkov Y. G. Methods for functional killing radioelectronics, Patent of Russian Federation 2154839, G01S13/00, 2000.

24. Kopp C. The e-bomb—A weapon of electronical mass destruction. In *Information Warfare*, Schwartau, W., ed. Thunder's Month Press, New York, 1996.

25. Gurevich V. I. Electromagnetic terrorism—The new reality of the 21st century. *The World of Technics and Technologies*, 12, 2005, 14–15.

26. Prishchepenko A. B. The weapon of unique possibilities. *Independent Military Review*, (26), July 17–23, 1998.

27. Kadukov A. E., Razumov A. V. The basics of technical and prestrategic application of electromagnetic weapon. *The Journal of Electronics of St.-Petersburg*, 2, 2000.

28. Sokot S. Russia markets the weapon of future. *Independent Military Review*, 39 (261), October 19–25, 2001.

29. Prishchipenko A. A new challenge of terrorists—Electromagnetic. *Independent Military Review*, November 5, 2004.

30. Bogdanov V. N., Zhukovskiy M. I., Safronov N. B. The electromagnetic terrorism—The state of the problem. Presented by the Scientific and Technical center "Atlas" of FSS of Russia. *Proceedings of the Conference "Informational Safety of Regions in Russia – 2005,"* St. Petersburg, Russia, June 14–16, 2005.

31. Daamen D. Avant-garde terrorism: Intentional electro magnetic interference. On methods and their possible impact. Report. Spring 2002.

32. Gannota A. The object of defeat—Electronics. *Independent Military Review*, 13, 2001.

33. Gazizova T. R., ed. *Electromagnetic Terrorism at the Edge of Centuries*. Tomsk State University, Tomsk, Russia, 2002.

34. Dobrykin V. D., Kupriyanov A. I., Ponomaryov V. G., Shustov L. N. *Radio-Electronic Battle: Forced Defeat of Radio-Electronic Systems*. Vuzovskaya Kniga, Moscow, Russia, 2007.

4

Vulnerability of Modern Relay Protection to Cyber Attacks

4.1 Dangerous Tendencies

Modern trends in relay protection (RP) based on the substitution of electromechanical protection relays (EMRs) by digital protection relays (DPRs) have resulted in the emergence of an absolutely new problem, which was not known before. This problem is the possibility of an intentional remote destructive impact (IRDI) on RP in order to put it out of action or make it perform functions that have nothing to do with the current operational mode of protected electrical equipment. In the modern power systems, DPR is the most critical link [1], in which on the one hand it is the most susceptible to IRDI, while on the other hand it is directly connected to a circuit breaker influencing the state of the power system. This is why the IRDI in the form of cyber attacks [2] and intentional electromagnetic destructive impact (EMDI) [3,4] are aimed initially at DPRs.

Special research conducted by the B5 committee of CIGRE and presented in its report confirmed the relevance of the problem and the conclusion that expansion of application of the most advanced standard International Electrotechnical Commission (IEC) 61850 with its GOOSE messages as well as modern Ethernet technologies in RP results in increasing its susceptibility to cyber attacks [5,6].

The appreciation of the problem of DPR's cyber safety has resulted in intensification of multiple investigations in the area of cyber safety all over the world. For example, in the United States, this situation is dealt with by a large department, which consists of several thousands of people under the supervision of the Head of National Security Agency (NSA) General A. Keith [2], while in Russia, this job is performed by a special department of the Federal Security Service of Russia. There are also edicts of the President of the Russian Federation: "About establishment of a state system of detection, prevention and liquidation of consequences of computer attacks on informational resources of RF" and "The framework of state policy of RF in the area of international information security by 2020," which are viewed as a reply to the "International Strategy for Cyberspace" adopted by the United States in 2011 [7]. It is the first time that the United States sets computer subversions on a par with traditional military actions, reserving the right to react with all available means, up to the use of nuclear weapons. It is known from the mass media that the problem of cyber security of the Israeli power system is handled by a special department of Israel Security Agency (SHABAK) together with the specialists of the Israel Electric Company. In 2011, Israel launched National Cyber Command (the National Cybernetic Taskforce) into the Administration for the *Development of Weapons and Technological Infrastructure* (*MAFAT*) for coordination and concentration cyber R&D activities for the various branches of the Israeli defense establishment (cyber departments of Mossad, SHABAK, Ministry of Defense (Unit8200), and other entities) with a budget of hundreds of millions of shekels in a year. Recently, the major Russian institute—All-Russian Research and Development Institute of Relaying (VNIIR)—also established a special department, which is dealing with specific problems of cyber security of RP.

Last year, the U.S. Cyber Command started its operations. Being a part of the NSA, the top secret and one of the most powerful intelligent agencies of the world, under the command of General *Keith Alexander*, the organization has united all previously existing cyber safety departments of the Pentagon. A year ago, Cyber Command had about 1,000 employees, but the military had announced its initiation of a major hiring plan for particular specialists to increase the number of employees of this unit of NSA to 10,000 people. Some of them will be in charge of the protection of military and state infrastructures as well as of the most important commercial properties of the state. It's obvious that such a large structure will not only protect from the hacker attacks but also develop hacker attacks (after all, attack is the best defense). Top-secret documents reveal that the NSA is dramatically expanding its ability to covertly hack into computers on a mass scale by using automated systems that reduce the level of human oversight in the process. The classified files—provided previously by NSA whistleblower Edward Snowden—contain new details about groundbreaking surveillance technology the agency has developed to infect potentially millions of computers worldwide with malware "implants" [8].

The current Head of Cyber Command and Director of the NSA, General K. Alexander, has declared at hearings of the Military Service Committee of House of the United States that the effect of cyber weapons is comparable to the effect of mass destruction weapons.

Cyber weapons are on the fast track. Experts say that many countries—including the United States, Russia, China, Israel, Great Britain, Pakistan, India, and North and South Korea—have developed sophisticated cyber weapons that can repeatedly get into computer networks and are capable of destroying them. In 2010, the cyber budget of the United States reached 8 billion USD, and in the future, this amount will further grow. In 2011, the United States plans to embrace a new doctrine of cyber safety. Its aims were revealed in a policy paper by Deputy Head of the Pentagon William Linn III under the symbolic name "New Space Protection." The main idea of the paper is that from now on, the United States considers cyberspace as the potential battlefield along with land, sea, and air. In parallel, NATO has started to work on the development of a collective cyber safety concept. At an Alliance Summit held in November 2010, it was decided to develop a Cyber Safety Action Plan. The document should be ready by April 2011 and signed in June. The main concept of the document centers around the creation of a NATO Cyber Accident Response Center. Initially, the center was planned to be commissioned in 2015, but on the insistence of the United States, this has been reduced by 3 years.

According to Gartner, the volume of the international market of cyber security increased from 61.8 billion USD in 2012 to 67.2 billion USD in 2013. It is expected that it will reach 86 billion USD by 2016.

Effectiveness of cyber weapon was proved by the widely known cyber attack on the uranium enrichment plant in Natanz, Iran, with Win32/Stuxnet worm that destroyed hundreds of centrifuges.

Another massive attack was mounted in September 2011 on the Japanese corporation Mitsubishi Heavy Industries, engaged in the production of the important parts of the F-15 jets, Patriot missile complexes, submarines, surface ships, rocket engines, guidance and intercept ballistic missiles systems, and other military equipment. The computer equipment of the corporation (45 private servers and around 50 PCs) was infected with a set of viruses, which took complete control over it. The viruses allowed controlling the computers from the outside and transfer the available data. Some viruses were aimed to activate the built-in computer microphones and cameras. This allowed the attackers to keep an eye on what's happening in the production and research facilities. Other viruses erased signs of cracking, which seriously complicated the assessment of the damage. Information from the computers was transferred to 14 websites located in other countries, including China, Hong Kong, the United States, and India.

Modern technologies allow infecting the computer system remotely through the encoded radio signals sent by unmanned aerial retransmitters. Wi-Fi systems, which are planned to be the basis for smart grid, are particularly exposed to such infections. Built-in Wi-Fi modems have already been built into the DPRs by the leading Western manufacturers.

In the past, there were attempts of computer penetration into the power system of Israel made by Iran. Senior CIA analyst Tom Donahue at a meeting between government officials and employees of U.S. companies from the power, water, oil, and gas supply sector mentioned that CIA identified numerous attempts to penetrate the U.S. power grid. Obviously, we can state that the cyberwars have already started and while they will intensify over time, the vulnerability of power systems will continue to increase too due to the current tendencies, thus forming a very dangerous vector, especially in RP as a very vulnerable part of the power system [9].

It is known that protection relay has two types of failures: the so-called "nonoperation" and the "unnecessary operation" (which in this context is similar to faulty operation). As mentioned in Ref. [1], the unnecessary (faulty) operation of RP can result in more significant damages than nonoperation. This is due to the fact that nonoperation of protection of a certain type is backed up by protection of other types or more remote protection, such as protection on other levels, while the unnecessary operation of RP is almost impossible to prevent by available means. This idea is not something unexpected and has been mentioned before elsewhere [10]. At the same time, Ref. [1] mentions that this is an absolutely different situation, when an inadequately operating protection relay can generate a command for switching off a circuit breaker in the case of unnecessary operation and thus artificially prevent a power system from functioning normally. This not only leads to the disconnection of thousands of consumers and considerable damage comparable to emergency mode in a power system but also creates the danger of a serious blackout caused by sudden overflows of power in the case of disconnection in a branched energy supply system. As mentioned in Ref. [11], 25%–28% of significant blackouts known in the world were the result of protection relay failures. If we admit that 50%–70% of the transition of an ordinary emergency mode to a serious blackout is also caused by protection relays [11], we can conclude that protection relay is responsible for all the blackouts. ORGRES representatives (Moscow) presented interesting data, which confirm the aforementioned idea [12]:

> In 2012 there were 53,214 events of actuation of protection relays and automatic equipment on Federal Grid Company's equipment. This includes 52,763 events (99,15%) of correct actuation; 451 events of incorrect actuations, which *include 213 events of unnecessary actuation, 160 faulty actuations and 76 failures to actuate.* …
>
> The index of correct operation of DPR in 2012 was 98.97%, which is lower than the generalized index of correct operation of electromechanical protection relays (99.31%).

First of all, this means that the number of faulty and unnecessary actuations (373) under general operational conditions (i.e., without IRDIs) is much higher than the number of nonoperation (76). Secondly, it suggests that the reliability of modern DPRs is lower than that of old and worn-out EMRs.

4.2 Cyber Security

This being said, let us clarify what "cyber security" means. Analysis of several related publications shows that this term usually means informational security. It should be considered that "information security" may mean different things in different contexts; such meaning can be wider or narrower. The wider meaning includes the whole spectrum of organizational and technical measures of security provision. The types of information security are provisionally divided into passive and active. The passive risk of information security is aimed at illegal use of information resources and is not aimed at setting the information system out of order. This type of risk includes access to databases or listening through the data transfer channels. The active risk of information security is aimed at setting the information system out of order by an intentional attack on its components. The active threats of computer security include physically knocking the computer out of operation or disturbance of its performance as well as intentional interference with the normal mode of operation of equipment controlled by the computer by interference with the algorithm of its operation. A typical example of the latter would be a well-known virus called Stuxnet [2]. Under information security, I will mean the ways of information protection from intentional or accidental unauthorized access, which can damage the normal course of data exchange in a system as well as stealing, modification, and destruction of information. By changing the programmable logic controller (PLC) code, the virus tries to reprogram controllers of industrial systems (in particular, Siemens controllers) to get control of them under the table. According to some data, every day this virus makes several thousands of attacks on Siemens controllers. We think that this should be noticed by the technical community and first of all by power industry managers filled with admiration for mastering such technologies as smart grid that is widely implemented in some countries under a government program. Moreover, this dangerous virus has already infected controllers of power systems! In 2009, the U.S. government admitted the detection of a virus capable of disabling power objects of the country.

The real issue is the jump of the smart grid vulnerability to hacker attacks. In fact, if all elements of the smart grid are controlled by commands through the networks with TCP/IP protocols, there is a huge risk of external intervention to the power system operation. Many experts emphasized this hazard devoting international conferences to it. Only apologists of the smart grid, for some reason, "do not notice" these problems. What do we hear from the apologists of the smart grid? Nothing but usual reservations about the necessity to isolate the internal network of the smart grid from the external web (this concept was realized in Iran), about access passwords and other trivial safety measures. We all understand that all these measures can limit access for normal people, but not for experienced hackers cracking even the very well-protected networks of the Ministries of Defense and banks.

However, hackers are not the major concern since the armies of many countries of the world have special divisions consisting of skilled professionals intended for cyberwars, that is, for cracking and sabotaging the protected computer networks of the enemy. It is safe to say that the computer network of the smart grid will be the number one target for such divisions. "Welcome to 21st century war," says Richard A. Clarke, former Special Adviser to the

U.S. President George Bush for Cyber Security and National Coordinator for Security and Counterterrorism, "Imagine the bursting electric generators, the derailing of trains, the crashing of planes, the blowing up of gas pipelines, the arms systems suddenly ceasing to work, and armies which do not know where to move." This is not an episode from the next Hollywood blockbuster; it is a summary of the consequences of the new type of battle conducted as cyberwar by skilled American experts.

The major problems that need to be solved in the area of engineer-technical protection of information include

1. Interception of electronic emanation and electric signals
2. Forced electromagnetic irradiation (lighting) of communication lines in order to obtain a parasitical modulation of the carrier
   a. Implementation of listening devices
   b. Remote photography
   c. Interception of acoustic irradiation and restoration of text sent to a printer
   d. The copying of information carriers, breaking through protection
   e. Impersonation
   f. Masking under system's queries
   g. Use of software traps
   h. Use of drawbacks of programming languages and operating systems
   i. Unauthorized connection to hardware and communication lines of specific devices, which provide access to information
   j. Malicious setting of protection mechanisms out of order
   k. Deciphering of encrypted information by special software
   l. Informational infections, that is, different viruses, including "logical bombs," "Trojans," "worms," and "password interceptors"

In order to ensure information security, the following measures are usually implemented:

1. Firewall—a complex of hardware or software measures, which control and filter network packets flowing through it in accordance with the established rules. This means enables
   a. Filtering access to initially unprotected services
   b. Preventing in obtaining closed information from protected subnetwork as well as intrusion into a protected subnetwork of faulty data by means of susceptible services
   c. Control access to network nodes

    d. Registering all access attempts from both external and internal networks

    e. Regulating the order of access to the network

    f. Notification of suspicious activity, flexing, or attacks to network nodes or the firewall itself

2. Antivirus software developed to locate computer viruses as well as malicious software and restore the infected (modified) files and to prevent infecting files and/or the operating system. Location of viruses is usually based on comparison of codes browsed by the antivirus with the known codes (signature) of malicious software set up in the library of the antivirus. Recently, the so-called proactive technologies of antivirus protection have started to develop. The idea behind them is that unlike reactive (signature-based) technologies, they prevent infection of the system rather than searching for malicious software in the system.

3. Cryptographic methods of protection of information, in other words coding and encryption of information, access keys, special protocols of network, and user authentication.

These widely known technical measures can be supplemented by some specific measures accepted in digital RP. One of these measures is a use of general information data buses (process buses) since the attack on such a bus is the simplest and the most efficient way, which can interfere with operation of a substation. It is possible to use several "point-to-point" links instead of these buses. This will allow using commutation protocols (including one-way data transfer), which are more resistant to attacks. These and many other specific measures of RP protection, which have more to do with protection of data transfer protocols, increasing of password cryptographic robustness, etc., are discussed in more details in Ref. [6].

4.3 Are Widely Known Measures of Information Security Enough to Ensure Reliable Operation of Digital Protective Relays?

The main question now is, "Are all of these widely known measures of information security enough to ensure reliable operation of DPR?" My answer is, "No." Traditional and well-known methods ensuring information safety cannot fully prevent unauthorized actions of RP. It doesn't mean that some methods of protection are not efficient enough yet (which is actually the case), but it means that it is not possible in principal. The analysis discussed earlier suggests that all the known technical measures of protection of information are designed to protect information channels from unauthorized access and information itself from being stolen and/or damaged. Of course, these information channels are widely used in digital RP, and they should definitely be protected very well. But here's the question: are these channels the only way to make DPR disconnect the circuit breakers and ruin the circuit? In point of fact, DPR contains a lot of so-called "logic inputs" (LIs) that are sensitive to voltage availability. This voltage is delivered to LI by means of contacts of external electromechanical relays. It is not possible to encrypt or encode the fact of voltage presence or absence on the LI. Moreover, the LI's design in a DPR is not suitable to receive encoded information. It is enough to modify the freely programmable logics of DPR so that during a remote supply of voltage by means of a certain external relay to a previously selected LI, there will be actuation of output DPR relays affecting the circuit breakers, and it will be possible to use it to sabotage the power system. Unfortunately, none of the preceding measures of protection from cyber attacks will help in this situation since in actuality there was no cyber attack to DPR.

Another example of low efficiency of known methods of protection from cyber attacks is a special virus that seizes computer networks and resembles the Win32/Stuxnet virus that defeated protected computer networks of the nuclear power program of Iran in September 2010. Win32/Stuxnet poses a threat to the power and industrial enterprises. During start-up, this malicious program employs the previously unknown vulnerability of USB-stick LNK files. Execution of the malicious code results from the vulnerability of the Windows Shell related to the display of dedicated LNK files. The new distribution method may open the door to other vicious programs that will use the same route since the vulnerability still exists. Win32/Stuxnet can also bypass host intrusion prevention system (HIPS) protecting the systems from external impacts as this malicious program contains files with legal signatures. Now this virus makes several thousands of attacks per day on the computers with Siemens installations. According to the experts, the attacks are directed to sophisticated systems such as automatic programs operating whole plants and municipal infrastructure units, including public water supplies. In published comments of analysts, they consider the attacks to Siemens devices as the first occurrence of mass "industrial sabotage." The complex analysis performed by the experts of Symantec showed that Stuxnet is an extremely dangerous and complex threat to the safety of the computer systems and is focused on the infection of industrial equipment monitoring systems that are also widely used at electric power stations. By changing the PLC code, the virus attempts to reprogram the industrial control systems (ICSs) in order to take over the control without operators being aware. The complexity of the virus and its extremely high selectivity testify that this malicious program was created by a group of the highly skilled experts possessing the huge budget and integration capabilities rather than by some self-taught hacker. After analysis of the code of a worm, the experts of Kaspersky Laboratory concluded that the Stuxnet "was not intended for espionage on the infected systems, it was developed for sabotage." "Stuxnet does not steal money, confidential information or send spam," Eugene Kaspersky confirms, "this thread is created to control productions, and literally to take over the operation of the production capacities. Quite recently we have struggled with cyber criminals and Internet hooligans, but now, I am afraid, the time has come of cyber terrorism, the cyber weapon and cyber wars."

Recently, there have been a lot of articles on the new reality of cyberwars in cyberspace published on the Internet and in the mass media, some of which are quite inflammatory. But from all publications, I wish to mark one very important document [13] (Figure 4.1) in which problem of cyber and electromagnetic security is incorporated in the common problem via the integration and synchronization of cyberspace operations and electronic warfare (as well as in the given book), requiring the complex solution. It is very true and important, tendency, in my opinion.

FIGURE 4.1
FM3-38. Cyber electromagnetic activities document.

References

1. Gurevich V. The issues of philosophy in relay protection. *World of Technics and Technologies*, 1, 2013, 56–58.
2. Gurevich V. Cyber weapons against the power industry. *Energize*, 10, 2011, 40–42.
3. Gurevich V. I. Stability of microprocessor relay protection and automation systems against intentional destructive electromagnetic impacts. *Electrical Engineering & Electromechanics*, 5 (Part I), 2011, 23–28.
4. Gurevich V. I. Stability of microprocessor relay protection and automation systems against intentional destructive electromagnetic impacts. *Electrical Engineering & Electromechanics*, 6 (Part II), 2011, 21–28.
5. The impact of implementing cyber security requirements using IEC 61850. CIGRE Working Group the B5.38, August 2010, 91pp.
6. Ward S., O'Brien J., Beresh B., Benmouyal G. et al. Cyber security issues for protective relays; C1 Working Group Members of Power System Relaying Committee. *IEEE Power Engineering Society General Meeting*, Tampa, FL, June 24–28, 2007, pp. 1–8.
7. White House. International strategy for cyberspace: Prosperity, security, and openness in a networked world. Sealed by the President of the United States, May 2011, 25pp. http://www.whitehouse.gov/sites/default/files/rss_viewer/international_strategy_for_cyberspace.pdf.
8. Gallagher R., Greenwald G. How the USA plans to infect 'millions' of computers with malware. *The Intercept*, March 12, 2014. http://fortunascorner.com/2014/03/12/glenn-greenwald-at-it-again-how-nsa-plans-to-infect-millions-of-computers-with-malware/.
9. INEEL/EXT-04-02428. A comparison of electrical sector cyber security standards and guidelines. U.S. Department of Homeland Security under DOE Idaho Operations Office, Idaho Falls, ID, October 28, 2004, 24pp.
10. Shalin A. I., Trofimov A. S. Efficiency and reliability of a relay protection. *Relay Protection and Substation Automation of Modern Power Systems, CIGRE-2007*, Cheboksary, Russia, September 9–13, 2007.
11. Saratova N. Y. The analysis of approaches to research of processes of system collapses. System researches in power. *Materials of Conference of Young Scientists*, Irkutsk, Russia, 2007, pp. 31–39.
12. Kuzmichev V. A., Sakharov S. N. Analysis of functioning of relay protection devices in the ENES in 2012 year. Theses of reports, Relavexpo-2013, Cheboksary, Russia, pp. 56–57.
13. FM3-38. Cyber electromagnetic activities. Department of the Army, Washington, DC, February 12, 2014.

5

Reducing the Vulnerability of Digital Protective Relays to Intentional Remote Destructive Impacts

5.1 Passive Methods for Protection against Intentional Destructive Electromagnetic Impacts

Ideal protection against intentional destructive electromagnetic impacts (IDEIs) would be the full isolation of electronics against the environment and covering the building with a bulk thick-walled ferromagnetic shield. At the same time, we must realize that, in practice, such digital protective relay (DPR) protection is impossible.

Thus, in practice, we have to use less reliable protection measures, such as conducting grids or conducting coating films for windows, honeycomb metal structures of air intake and air holes, as well as special conductive lubrication and conductive rubber gaskets located on the frames of doors and hatches.

Today, there are special metal cabinets (Figure 5.1) available on the market that ensure significant attenuation of IDEI. Standard cabinets made of iron sheets having no windows or gaps provide significant attenuation of IDEI. Galvanized assembling panels of such cabinets, as well as special conductive seals, significantly increase the effectiveness of such cabinets since galvanizing allows equalizing potentials within large areas (steel-specific resistance is 0.103–0.204 $\Omega \times$ mm^2/m, and zinc-specific resistance is 0.053–0.062 $\Omega \times$ mm^2/m). Aluminum has even lower resistance (0.028 $\Omega \times$ mm^2/m). Thus, some manufacturers produce single-block cabinets from a special alloy: Aluzinc 150 (Aluzinc®—registered trademark of Arcelor)—of which 55% of the material is covered with aluminum, 43.4% with zinc, and 1.6% with silicon. The surface of the cabinet with this covering provides a high deflection of IDEI.

These cabinets are manufactured and supplied to many countries by the Sarel company (today, Schneider Electric Ltd., Great Britain). Similar cabinets providing protection against IDEI are also manufactured by other

FIGURE 5.1
Control cabinet with upgraded protection against IDEI equipped with special loops, conductive rubber gasket, special coupling and connecting elements, shielded air vent windows, etc. (Equipto Electronics Corp.).

companies, such as Canovate Group; RF Installations, Inc.; Universal Shielding Corp.; Eldon; Equipto Electronics Corp.; ATOS; MFB; European EMC Products Ltd. AMCO Engineering; and Addison. This equipment usually attenuates the emission per 80–90 dB on a frequency of 100 kHz to 1 GHz.

Certainly, control cables must be shielded with twisted pair. The minimum requirement to the shield is high density of armor (not less than 85%). Double-shielded cables have a much better shielding effect (see Figure 5.2). For relatively low frequencies (up to several tens of MHz), the braided screen provides better shielding than the foil mainly due to its thickness. However, the shielding properties of the braided screen sharply decrease and become almost unacceptable before the frequency reaches 100 MHz. At the same time, the foil has flat

(a)   (b)

FIGURE 5.2
Double-shielded cable. (a) With double-braided screen; (b) with double combined shield (braided screen and foil).

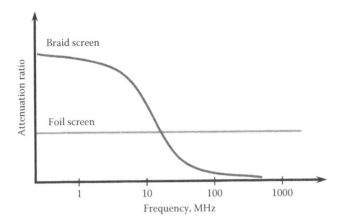

FIGURE 5.3
Dependence of shield factor from frequency for braid–foil shields.

FIGURE 5.4
Cable RE-2X(ST)2Y(Z)Y PiMF characterized as interference superstable (transmitting analog and digital signals up to 200 kbit/s; twisted-pair wires, each pair shielded with polyethylene (PE) foil; three-layer common foil shield armored with steel wire; external cross-linked polyethylene (XPLE) insulation; up to 24 pairs of wires per cable; can be used outdoors and for burial; has high mechanical strength).

amplitude-frequency response (AFR) maintaining acceptable shielding properties over a wide range of frequencies up to the GHz range (see Figure 5.3). Thus, cables with combined braid–foil shield are the most preferable. Excellent protection against IDEI uses cables combining twisted-pair wires, foil shields for each pair of wires, and three-layer common shield made of foil (see Figure 5.4).

The Belden Company has developed and patented a simple and effective method for shielding cables based on foil-coated PE film (polysandwich) under the name of Beldfoil®. The company produces cables with two layers of foil and braid, or even four layers, where foil interstratifies with braid two times combining best properties of foil and braid in one cable (see Figure 5.5).

The effectiveness of cable shielding depends heavily on the grounding effectiveness. As shown in Ref. [1], on the one hand, the grounding of control cable shield is effective only against capacitive pickups (referred to as electrostatic protection) and doesn't protect against inductive pickups (interference reduction factor $k = 1$) since the shield doesn't provide a chain for closing the interference current.

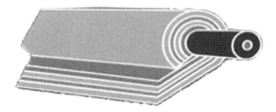

FIGURE 5.5
Multilayered shielding of polysandwich developed by Belden.

If the shield is two-side grounded, there is an additional chain (shield) with much lower impedance for high-frequency signals than the ground. As a result, the operating signal is divided into two components: low-frequency component goes through the ground and high-frequency component goes through cable shield. Therefore, for the high-frequency component, the current in the shield is equal to the current in the central core directed in the opposite direction and is compensated due to inductive coupling between the shield and central core. This provides protection against high-frequency pulses emitted from the central core to the environment (to adjacent cables) with an interference reduction factor $k = 3–20$. This system is also effective under an external electromagnetic pulse to the shield when the high-frequency signal induced into the shield is bridged through the ground. When connecting the shield to the ground bus, it should be considered that a "wrapping" connecting wire on the shield is unacceptable as well as coiling a long connecting wire between the shield and ground bus. Each additional loop increases the impedance of the grounding on high frequencies and significantly reduces its effectiveness. For cabling at substations, laying a two-side grounded potential-equalizing copper bus in parallel to the cable run can be an additional solution capable of improving efficiency of the shield. Its effect is provided by the fact that copper bus impedance on high frequencies is much less than ground impedance (and even shield impedance), so the main component of the pulse interference high-frequency current will run through the bus rather than through the shield.

While new cables with multilayered foil shielding are capable of effectively attenuating external IDEI, old types of cables with sparse braid do not satisfy these needs. In order to attenuate an external electromagnetic field, these old cables can be laid in metal trays and tubes. Plastic metalized trays widely used for laying control cables have the least shielding effect.

Due to a very thin conductive layer, such a structure operates effectively only on frequencies of 600 MHz and above. On frequencies under 200 MHz, it doesn't work at all [1]. At the same time, aluminum trays combined with copper cable braid can attenuate induced voltages 10-fold, and thus they can be widely used as effective electromagnetic pulse (EMP) protection measure. However, laying cables in steel water pipes ensures the best attenuation of induction over a wide range of frequencies.

Prevention of IDEI penetration into the apparatus through different cable entries and connections (plugs) is a more difficult technical problem than cable shielding.

Today, there are a lot of special connectors with integrated IDEI filters available on the market (see Figure 5.6), from many manufacturers, such as Amphenol; Spectrum Control Inc., Spectrum Advanced Specialty Products; EMP Connectors; ERNI Electronics; Sabritec; MPE; Glenair Inc.; Captor Corp.; and Lindgren-Rayproof.

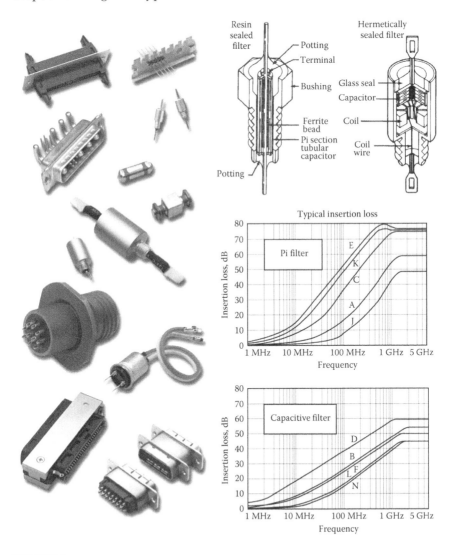

FIGURE 5.6
Several types of input connectors with integrated filters manufactured by Spectrum Control Inc.

As a rule, such filters are manufactured based on ferrite rings or combined inductances and capacitances (see Figure 5.7), installed into the connector (see Figure 5.8).

Filters, spark arresters, metal-oxide varistors, and Zener diode HS suppressants are widely used for protection of cable entries. The whole range of such devices is produced by the company RFI Corporation and others (see Figure 5.9).

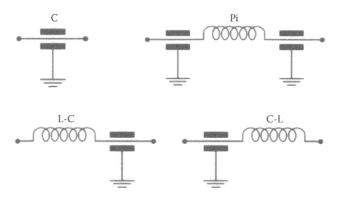

FIGURE 5.7
Typical circuits of filters integrated into connectors.

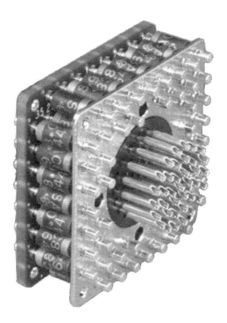

FIGURE 5.8
Design of a connector with integrated filters by Glenair Inc. A high number of filtering elements are installed between two plates.

FIGURE 5.9
Cable entries' filters manufactured by RFI Corporation.

FIGURE 5.10
Sealed power filters manufactured by Captor Corp. designed for AC and DC power circuits up to 100 A ensuring effective IDEI attenuation to not less than 100 dB within the frequency range of 14 kHz to 10 GHz.

The range of filters manufactured by this company includes filters for high currents (0.01–5000 A) and voltages (12 V DC to 5500 V AC).

Some manufacturers also produce power filters with wide frequency characteristics that are especially designed for protection against IDEI. Filters of the Captor Corp. demonstrate excellent characteristics (see Figure 5.10), and EPCOS power filters (see Figure 5.11), in the range of operating currents up to 150 A (surge currents up to 12 kA) and a voltage of 440 V. In such filters under the operating currents, the voltage drop reaches <1% per phase and attenuation reaches 100 dB over a frequency range of 14 kHz to 40 GHz. EPCOS also manufactures cabinet-type, three-phase filters working under the same frequency characteristics and operating currents of 1600 A, as well as low-power multichannel filters for actuating and control circuits.

Many manufacturers offer excess-voltage suppressors based on zinc-oxide varistors designed for 220/380/660 V circuits and allowing breakdown currents of up to 80 kA. Often, such devices contain, in series, short-circuit protection fuses protecting the circuit in case of varistor damage and a blown-fuse indicator (see Figure 5.12).

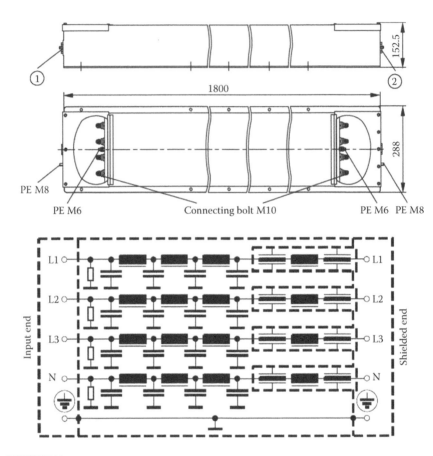

FIGURE 5.11
Dimensions and circuit diagram of three-phase EPCOS filter (150 A, 440 V).

Metal-oxide varistors have high power but not enough performance for protection against such kind of IDEI as high-altitude electromagnetic pulse (HEMP). Their parameters decline under the repeated high-power impulse loads. High-speed silicon Zener diode excess-voltage suppressors do not have such disadvantages (transient voltage suppressor [TVS] diodes). Their operation is based on a sharp drop of the resistance from relatively high value to almost zero under induced excessive voltage with a certain threshold (see Figure 5.12).

Besides, contrary to varistors, the parameters of such excess-voltage suppressors do not decline under the repeated high-voltage effects and mode switch (see Figure 5.13).

Unfortunately, most modern suppressors of this type have limited pulse power (up to 1500 W undervoltages of up to 600 V) and are suitable for protecting electronics inputs but not for power and supply circuits. However, several companies, such as Littelfuse, specialize in the development and

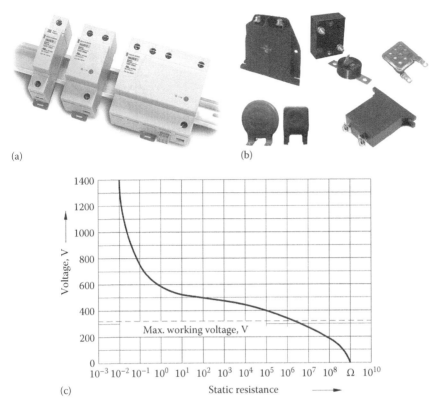

(a)					(b)

(c)

FIGURE 5.12
(a) High-capacity protecting devices based on metal-oxide varistors designed by Square D (Schneider Electric). (b) Powerful varistors of different types with rated voltage of 130–1100 V and breakdown current of 3–100 kA. (c) Typical volt–amp diagram of zinc-oxide varistors.

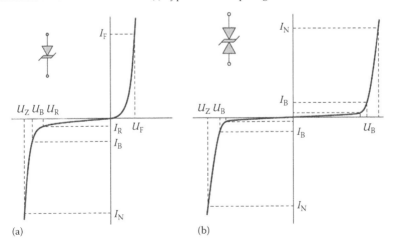

(a)					(b)

FIGURE 5.13
(a) Volt–amp diagram of monodirectional (DC) and (b) bidirectional (AC) diode suppressors.

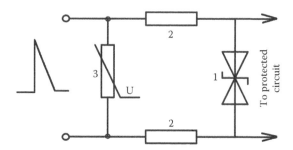

FIGURE 5.14
Hybrid protection device: (1) semiconducting suppressor, (2) current-limiting resistor, (3) powerful varistor.

production of elements protecting against surge voltage. Littelfuse, for example, manufactures suppressors of much higher impulse power up to 30 kW and discharge pulse currents up to several hundreds of amperes.

Varistors and transient voltage suppression (TVS) diodes can be connected in parallel in order to increase discharge current. In-parallel connection of different suppressors, such as varistors and semiconducting suppressors, enables improving efficiency of surge voltage protection (see Figure 5.14). Such a hybrid device demonstrates excellent characteristics: initial reaction is provided by fast-response suppressor 1 responding to pulse with an even steep leading edge and absorbing a part of its energy; discharge current is limited with resistors 2 preventing damage to suppressor.

The voltage drop on resistors 2 increases the voltage on varistor 3, resulting in a sharp decrease in its resistance and bridging resistors. The rest (the most part) of energy is absorbed with a powerful varistor.

While designing means to protect against intensive HEMP, it should be considered that only one type of protection is not capable of ensuring effective overall protection. Thus, only the combination of all available protection means can provide complete protection. One of such types of protection means is protection of buildings and premises against HEMP. The most effective protection is ensured with special panels combining EMP-reflecting and absorbing layers (see Figure 5.15).

However, fully shielded premises would cost a lot of money. Therefore, in practice, cheaper intermediate options including protective paints, films, curtains, and hangings can be used. Over recent years, significant progress has been made in developing conductive paintings and construction materials with unique properties and wide application, as well as clear conductive coatings that can be applied on the glass. Conductive paints, lacquers, and sprays based on copper, aluminum, brass, nickel, and graphite are manufactured by many companies, such as Caswell; YSHIELD EMR Protection Company; Less EMF, Inc.; Gold Touch, Inc.; Spraylat Corp.; Cybershield; Applied Coating Technologies Ltd.; and B.M. Industria Bergamasca Mobili S.P.A. High results show protecting

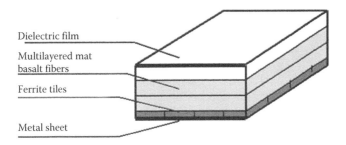

FIGURE 5.15
Integrated protecting panel "Ferrilar-5."

paint Tikolak developed by a Moscow company Tiko. Tikolak is a new patented (in Russia) universal nonmetal conductive coating material combining carbon filling compound with polymeric binding agent (8%–20% epoxy plus graphite-soot compound (11%–39%) with mass ratio of 0.1%:1.0%, hardener 0.5%–1.5%, organic solvent, etc.). According to Tiko, this coating ensures shielding against IDEI over a wide frequency range up to 300 GHz. Interior and exterior surfaces of a building coated with Tikolak are characterized with manifestly less IDEI penetrability. According to the manufacturer, one layer of Tikolak (only 70 μm) is able to reduce IDEI intensity by 3–3.5 times. This coating can be used on a variety of construction materials, such as chipboard, wood, gypsum board, as well as with any flexible material, such as fabric, leather, film, and paper. This coating can be covered with any decorating material, such as wallpaper, paint, and ceramic tile, and it costs much less than any foreign analogs (about $70 per kilo).

In order to get clear conductive glass reflecting IDEI, the oxide films of such metals as tin, indium, and zinc are used. Production of such glass is very complex and demanding while requiring costly equipment and qualified staff. Tiko has developed and patented (patent RF No. 2112076) a high-tech and economic way of covering the glass with conductive coating based on indium and stannum oxides. Clear conductive glass is manufactured by many companies, such as Tycon Technoglass; Pilkington; Shenzhen Wanyelong Industry Co., Ltd., and InkTec.

The Alfapol Company in St. Petersburg has developed construction materials based on shungite, which is a composite of solid carbon materials representing, in general, amorphous carbons close to graphite. The chemical composition of shungite is unstable: on the average, it contains 60%–70% of carbon and 30%–40% of soot. Soot contains 35%–50% of silicon oxide, 10%–25% of aluminum oxide, 4%–6% of potassium oxide, 1%–5% of sodium oxide, 1%–4% of titanium oxide, and other compound materials. Shungite combines the properties of regular construction materials with rather high electrical conductivity. This determines the ability to shield IDEI [2].

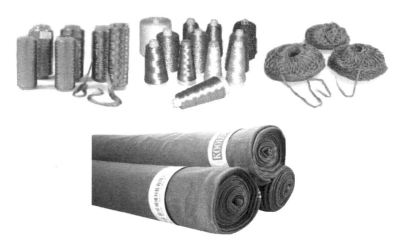

FIGURE 5.16
Conductive films, fiber, and fabric attenuating IDEI (up to 80 dB) manufactured by Koolon Fiber Tech. Corp.

According to Alfapol, shungite composite radio shielding materials can be divided in two classes by shielding method:

1. Construction materials, including concrete, bricks, and brick mortar. These materials are capable of providing IDEI energy attenuation at frequency ranges of more than 100 MHz at a level of not less than 100 dB. Their physical–mechanical characteristics match conventional construction materials. Shungite materials were tested in structures (concrete in slabs, bricks in blocking) and proved to be compliant with the current requirements.

2. Reconstruction materials, such as plasters and pastes for converting conventional premises into shielded. Layer of pastes (2–3 cm thick) provides shielding at a level of not less than 30 dB at a range of more than 30 MHz. Plaster composite "Alfapol SHT-1" provides attenuation of IDEI per 10–15 dB over a range of 10 kHz to 35 GHz with the thickness of plaster layer of 15 mm. Conductive curtains, fabric, and floor coating of different manufacturers can be used in addition to shungite walls (see Figure 5.16).

5.2 Improving Durability of DPR

In order to improve the durability of DPR, both technical improvements and organizational arrangements are required, in our opinion.

Technical improvement is to equip each DPR with a separate module containing special IDEI filters (ferrite rings, combination of different arresters, etc.).

All ingoing and outgoing DPR circuits should go through this module. All manufacturers of DPR should be obliged to equip their units with such modules. Such a module effectively matches the current module structure of DPR [3], and it can be replaced within the whole DPR lifecycle if new protection technologies and filtering modules appear in the market. This concept particularly includes implementation of standards for modular DPR construction and manufacturing DPR as standard modular boards that can be combined in relay protection (RP) cabinets with improved IDEI protection [3].

Today, Russian experts have investigated improving RP stability by implementation of two-level RP. Schedrikov proposes [4] effecting the first level of RP with DPR and the second level with a conventional electromechanical relay of type PT-40 plus a time relay of type PBM-12. Both sets of relays (DPR and electromechanical relay) are connected in parallel, and the electromechanical relay actuation time exceeds DPR actuation time by 0.1 s. Schedrikov expects that the electromechanical relay should spot for DPR in case of malfunction under emergency mode (in fact, it operates under the logic OR function). It should be noted that the in-parallel connection of DPR and electromechanical relay is not something unknown and has been practiced for a long time (see Figure 5.17) [5].

FIGURE 5.17
Distance protection of lines based on MPD type MiCOM P437 (bottom) and on electromechanical type LZ-31 (top) connected in parallel.

However, such a connection scheme doesn't eliminate false responses of DPR under IDEI, which can result in at least as serious power network problems as malfunction. In his next article, Schedrikov proposes changing from an in-parallel connection of electromechanical relay and DPR to the scheme where the electromechanical relay of type KPB-126 permits actuation of a breaker trip coil through microprocessor relay (using the logic AND function). Surely, such actuation ensures improved RP stability to false responses at IDEI but reduces total RP reliability (it is the inevitable consequence of improved sustainability to IDEI).

It is fair to say that we proposed to improve DPR sustainability to powerful intentional electromagnetic interference (EMI) with electromechanical relay permitting DPR actuation in Ref. [6] 15 years before the proposal of two-level RP was published and we suggested the idea about hybrid (mechanical semiconductor) RP device almost 20 years ago [7]. Moreover, it was not an abstract idea; rather it was a real design [8–10]. Today, thanks to the advent of a new element base, such as miniature high-voltage reed switches (RSs), RSs for large switched currents, small-sized transistors and thyristors with an operating voltage of 1200–1600 V, and switched currents of tens of amperes, there are new opportunities for the creation of hybrid relay (as a stand-alone protecting relay or as a starting unit [SU] for DPR).

In the publication [11], we attempted to demonstrate the schematic of modern hybrid relays. In the electromechanical part, we recommended using RSs.

Their distinctive features are high reliability (if standardized current and voltage limits are observed), fast response (fractions and units of milliseconds), excellent dust and moisture protection, no stripping and regulating during operation, small dimensions, full galvanic isolation of control circuit (coil) from output circuit (contacts), and possibility to obtain high-voltage isolation between the control circuit and the output circuit with very simple means [12]. Another specific example of such hybrid protective relay is fast-speed overcurrent relay, which we designed especially for separating network automation (see Figure 5.18) [13]. This device is very simple and contains a minimum number of elements selected with high voltage margin. Thus, for example, the thyristor is designed for 1200 V, and a miniature vacuum sealed switch is designed for 2000 V. Isolation between input coil and RS withstands voltages of 5 kV, which can be increased if necessary. A suppressor can be added connected in parallel to protecting varistor, as shown in Figure 5.14.

As mentioned earlier, recently, new hazards have appeared that encourage continuing to use electromechanical protective relays resistant to powerful IDEI. Conversely, new types of electromechanical relays capable of ensuring backup protection should be developed based on up-to-date technologies and materials.

Consequently, methods for improving DPR durability shouldn't include only technical innovations of the DPR structure. Organizational

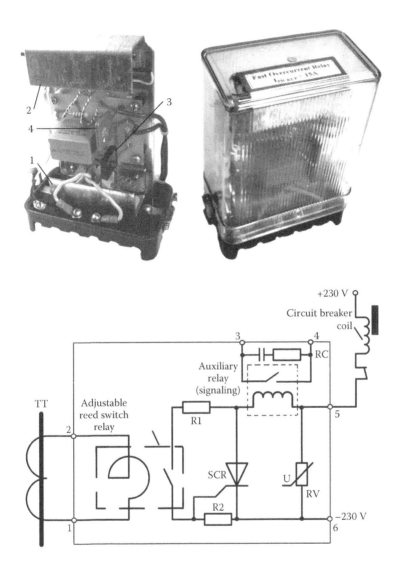

FIGURE 5.18
Fast-speed overcurrent hybrid relay. (1) Module of adjusted reed relay, (2) slave reed relay in ferromagnetic shield, (3) high-voltage thyristor, (4) varistor.

arrangements should include the stocking of printed board sets (modules) for DPR and proper storage. Since even disabled electronics can be damaged, such printed boards for DPRs should be stored in special well-shielded metal boxes. Modules of the central processor should be fully operational without the need of programming and setup. Since it is not possible to provide spare printed board sets for all DPRs used in power systems for economic reasons, the most critical DPRs of the power system should be determined in advance in order to have enough spare boards. For DPRs having no spare boards, correct removal methods should be considered. Substations and electric stations should have complete and adjusted sets of protection panels based on electromechanical relays that can be rapidly put into operation in case of mass problems with DPRs.

5.3 Active Method for Combined Protection of DPR against Cyber and Electromagnetic Threats

5.3.1 Device for Active DPR Protection

The recent appreciation of the problem of DPR's cyber safety has resulted in intensification of multiple investigations related mainly to sophistication of computer communication protocols designed for RP and improvement of their cryptographic security. Until recently, different specialists have concentrated all their efforts into just this area. As for IDEI, unfortunately this problem has not been seriously addressed yet. At the same time, 17 years ago when the DPR's problems were just emerging, the author offered a general idea of highly efficient compound protection of DPR from cyber attacks and IDEI by means of hardware facilities instead of software tools. The suggested protection device implements a principle of disconnecting and bypassing the sensitive DPR's terminals by means of responsive electromechanical RSs [6]. The idea of implementation of responsive electromechanical RSs was further developed in more details [14,15].

As mentioned before, the task of increasing the reliability of RP cannot be fulfilled by combining DPR's functions with those that have nothing to do with RP, such as monitoring of the functionality of electrical equipment and remote control of circuit breakers (RCCBs). The DPR should be used solely to solve problems of RP. Moreover, there are many specific devices in the market that can be used to solve other problems, such as the monitoring of electrical equipment. These devices may vary from the simplest relays that control the continuity of the circuit breaker (CB) trip coils to sophisticated complex units that ensure online control of gas composition dissolved in the transformer's oil or the level of partial discharges in the insulating material. As for RCCBs by means of DPR, this type of application will make it difficult to distinguish between authorized access and unauthorized access; this is why the use of this type of DPR should be eliminated. Moreover, with the separation of the functions, the hardware facilities ensure easier protection from intentional remote destructive impact (IRDI) also of RCCBs [16].

The general idea behind the suggested hardware-facilitated method of protection of DPR from IRDI is to use an electromechanical RS SU in combination with DPR and connected functionally in series with it as well as an electromechanical action element (RR1–RR7), which ensures the

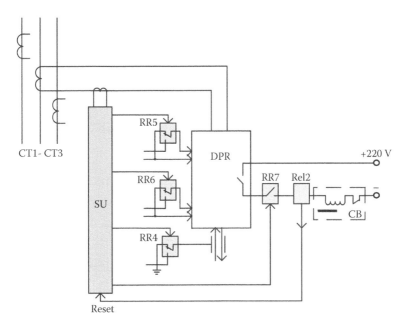

FIGURE 5.19
A structural diagram of DPR protection from IRDI.

blocking of the sensitive inputs of DPR and disconnection of its output circuit (Figure 5.19). The reset of the actuated SU is performed upon the CB's actuation and backed up by the RESET command at the end of a preliminary setup time period.

Without current and/or voltage actuation of this SU, DPR will not be able to influence the operation mode of the power system even under IRDI. If the SU is actuated and DPR enabled, nothing will interfere with using specific features and wide functional capabilities of DPR. At the same time, unnecessary actuation of the SU itself does not influence the operation of the RP, and thus there are no specific requirements as to the accuracy of the SU actuation. The only thing that is important is that it should always be actuated before DPR, that is, its settings should be a little bit lower than required for the controlled parameter. If the SU actuation was unnecessary and DPR was not actuated, the device would automatically reset. The main technical requirements for this device are its high reliability, insensitivity to short electromagnetic impulse (micro- and nanosecond range) and high-frequency interferences, resistance to substantial overvoltages, high level of galvanic insulation from external circuits, and high speed of response to actuation (several milliseconds).

The principle of operation of this device (Figure 5.20) is as follows. In its initial state under the normal operation mode of the protected object, all the input RSs (current and voltage sensors) RR1–RR3 are in the released state. The thyristor VT1 is in the off state; the control coils of the RSs RR4–RR7 are

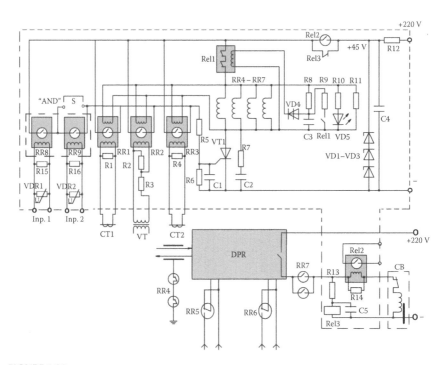

FIGURE 5.20
A diagram of improved device protecting DPR from IRDI.

not energized. The changeover contacts RR5 and RR6 short-circuit the logic inputs of DPR, the RR4 terminals short-circuit the communication channel, while the RR7 contacts disconnect the output circuit of DPR. Under these conditions, the DPR is fully blocked both in inputs and in outputs, and no IRDI can result in its unnecessary actuation and unauthorized actuation of the CB trip coil. Bypassing both the logic inputs of DPR and the communication channel also increases its operational vitality under the impact of a powerful electromagnetic impulse.

In case of the emergency mode in the protected object, at least one of the controlled parameters (current, voltage, or power) will drastically change. This change leads to actuation of at least one of the RSs RR1–RR3 within 1 ms or less. When actuated, an RS of a corresponding input starts vibrating at a doubled frequency. During the first event of the switching of the RS's terminals, the thyristor VT1 will switch on within several microseconds, and the control coils of the RSs RR4–RR7 will be powered. Actuation (opening) of RR4–RR6 RSs takes place during 2–4 ms, while the switching on of the power terminals of the RR7 RS (Bestact R15U RS type) does not take longer than 5 ms. Thus, the total response time of the unit to an emergency mode does not exceed 6 ms, which is quite acceptable considering the DPR's own actuation time of 30–40 ms. Under this mode of operation of DPR protection device, the DPR will be fully unblocked and returned into its normal mode of operation, retaining all its settings and features.

As can be seen in the diagram (Figure 5.19), each of the input relays (sensors) is equipped with a second winding on the RS, which receives power from the constant voltage source upon thyristor's VT1 switching on. Due to the additional magnetic field created by this winding, the RS of the actuated relay stops vibrating and enters a steady on state.

After the DPR performs the time delay setup by its feature, its internal output relay will energize trip coil of the CB via closed contacts RR7. Current flowing in the circuit of the CB trip coil results in actuation of the RS relay Rel2 with an addition powerful Bestact R15U RS and switching on of its terminals connected in parallel to the normally closed contacts of Rel3. Rel3 is actuated with a small time delay (about 10–20 ms). This time delay is necessary in order for Rel2 terminal to switch on before Rel3 terminal switch off.

At the end of actuation cycle of the CB, its interlock will switch off and the circuit of the trip coil will be interrupted. At the same time, Rel2 is released and its contact interrupts the anode circuit of the thyristor VT1, which is then immediately switched off, cutting off current from the control coils of RR4–RR7 relays and addition DC coils of RR1–RR3 relays. The device is totally returned to its initial state and is ready for a new cycle of operation.

If actuation of the device was unnecessary and DPR did not generate a command to disconnect the CB, the supply circuit of the thyristor VT1 will be interrupted for a short time by a normally closed contact of Rel1 relay upon the charging of C3 capacitor through resistor R8 and switch on of VD4 trigger diode (dynistor). The capacity of this capacitor and resistance of the resistor ensure a time delay of several seconds, which exceeds the maximum possible time necessary to fulfill the full cycle of DPR operation in order not to interfere with its operation should it be required. Actuation of Rel1 relay is temporary, since immediately upon its actuation and the opening of the normally closed contacts in the thyristor's circuit, its normally open terminal will switch on and discharge the C3 capacitor through a low-ohm R9 resistor, ensuring its full discharge and return into initial state. At the same time, the VD4 dynistor diode will close and current on the Rel1 relay's coil will be cut off. This is how a forced reset of the device to its initial state happens if its actuation is unnecessary.

The R11 resistor is needed to increase the current rate flowing through the power thyristor VT1 and its reliable maintenance in a conducting state. The LED VD4 serves as an indicator of the device's condition.

In order to increase the reliability of the device and its resistance to IDEI, only a few solid-state elements are used. They were selected with very big reserves for the device's maximal values in marginal current and voltage rates that are not used in usual industrial applications. For example, the VT1 thyristor with its actual operational voltage of 45 V was selected for maximum voltage rate of 1200 V; with its actual operational current rate of fractions of ampere, it can work under current of tens of amperes and conduct short-term impulses of hundreds of amperes. The Zener diodes VD1–VD3, as well as the VD4 dynistor diode, are also selected with very large power reserves. The auxiliary relays Rel1 and Rel3 are selected as sealed with high-power contacts.

The contacts of external relays intended for activation/deactivation of internal functions of DPR, determining a state of an RP, should connect and protect (bypassing in normal a mode) the logic inputs of DPR shown on Figure 5.20, also on addition logic inputs (Inp.1, Inp.2). Thus, for the purpose of increase of security of the system, not less than two input signals for single event, acting from two sources, are necessary. Inside of the device, these signals are galvanically isolated from external circuits by additional reed relays RR8 and RR9 by means of which logic function "AND." As a result, the activation of the device and deblocking of all logic inputs for the performance of all necessary operations is realized. It is not necessary that both these sources should be in the form of discrete signals (as contacts of the external auxiliary relays). One of them can be a discrete signal and another analog, in the form of a current or voltage acting on corresponding input of reed relay. The contacts of these relays connected in series (the logic function "AND"). Thus, the output of the reed relay of the second (not used) logic input can be bypassed by crosspiece S.

Stability of these additional logic inputs to short pulse noise even with very high amplitude provided with a corresponding level isolation between a coil and RS and its own time delay (it cannot switch on in time smaller than several milliseconds and thus is natural and very effective filter of high frequency and pulse noises).

For high noise stability to any sort of transients in auxiliary DC power supply, these additional inputs should have resistance much lower than the usual DPR logic inputs that is reached by using resistors R15 and R16.

Though coils of reed relays RR8 and RR9 are much stable to impulse over-voltage than semiconductor elements, in the device, measures are taken for its addition protection against the impulse overvoltage arising at IDEI, by means of varistors VDR1 and VDR2.

5.3.2 General Recommendations for Selection of Hardware Components of Protection Device

Miniature vacuum RSs with withstanding voltage of not less than 1 kV and having their own operate time of about 1 ms (Table 5.1) can be used as sensitive threshold elements in the SU (Table 5.1).

An assembly including an RS and a current coil with several winds of a thick wire is located into a ferromagnetic screen.

Table 5.2 shows the parameters of several types of thyristors that are more suitable to be used in a SU. In order to reduce noise resistance of a SU, it includes additional RC elements.

TABLE 5.1

Main Parameters of Vacuum Normally Open High-Voltage RSs of Several Types

Parameter/Type	MRA5650G	KSK-1A75	HYR2016	HYR1559	MARR-5	KSK-1A85
Contact type	NO	NO	NO	NO	NO	NO
Switching voltage (V)	1000	1000	1000	1500	1000	1000
Switching current (A)	1	0.5	1	0.5	0.5	1
Switching power (W)	100	10	25	10	10	100
Dielectric strength (V)	1500	1500	2500	1500	2000	4000
Operate time (ms)	0.6	0.5	0.8	0.4	0.75	1.0
Release time (ms)	0.05	0.1	0.3	0.2	0.3	0.1
Balloon dimensions	$D = 2.75$ $L = 21$	$D = 2.3$ $L = 14.2$	$D = 2.6$ $L = 21$	$D = 2.3$ $L = 14.2$	$D = 2.66$ $L = 19.7$	$D = 2.75$ $L = 21$
Sensitivities (A)	20–60	15–40	15–70	15–50	17–38	20–60

Note: NO, normally open.

Since the total current consumed by RR4–RR7 windings of a final control relay can be less than the latching current (I_L) and the holding current (I_H) of thyristor, the circuit in Figure 5.3 is supplemented by a powerful resistor R11, which increases the total current flowing through the thyristor up to 250–300 mA. Though there are special thyristors with increased sensitivity and small latching and holding current rates, which do not exceed 10 mA in the market (TS820-600, TIC106, BT258-600R, X0402MF, MCR708A1, etc.), we do not recommend to use them in this device since it can lead to reduction of its noise resistance.

Gas-filled RSs Bestact R15U by Yaskawa Company, which block output contact (trip contact) of DPR, can be successfully used as contact of final control relay (Figure 5.21); they are intended for current rates of up to 30 A at 240 V and have operate time, which does not exceed 5 ms.

High-voltage vacuum changeover RSs of different types (Table 5.3) can be used in a SU for shunting of sensitive (noncurrent) inputs of DPR.

Obviously, according to the theory of reliability, the connection of additional contacts (even those that are highly reliable) in series with trip terminals of DPR or parallel to its inputs will result in a certain reduction of reliability of RP. How much? It is very difficult to answer this question today, as there is no adequate information due to lack of experience of using such devices. However, should it be necessary, this reduction of reliability can be simply compensated by using two RSs connected in series or parallel as additional contacts (Figure 5.20). The probability of such failures as "unwanted operation" in electromechanical relays is much lower than the probability of "nonoperation." That's why their parallel connection (unlike ordinary parallel connection of DPR) will definitely increase reliability of RP. For normally

TABLE 5.2

Most Important Parameters of Several Types of Powerful Thyristors Intended for Mounting on a Printed Circuit Board

Parameter / Type	CLA50E1200HB	25TTS12	30TPS12 30TPS16	BTW68–1200
V_{RRM}/V_{DRM} (V)	1200	1200	1200	1200
			1600	
$I_{T (RMS)}$ (A)	79	25	30	30
$I_{T (AV)}$ (A)	50	16	20	19
I_{TSM} (A)	650	300	250	400
I_{GT} (mA)	50	60	45	50
I_L, max (mA)	125	200	200	40
I_H, max (mA)	100	100	100	75
dv/dt (V/µs)	1000	500	500	250
T_{GT} (µs)	2	0.9	0.9	100
T_j (°C)	−40 +150	−40 +125	−40 +125	−40 +125
Case type	TO-247	TO-220AC	TO-247AC	TOP3 ins.

Parameter / Type	CS 20-12io1 CS 20-14io1 CS 20-16io1	CS 30-12io1 CS 30-14io1 CS 30-16io1	CS 45-12io1 CS 45-16io1
V_{RRM}/V_{DRM} (V)	1200	1200	1200
	1400	1400	1600
	1600	1600	
$I_{T (RMS)}$ (A)	30	49	75
$I_{T (AV)}$ (A)	19	31	48
I_{TSM} (A)	200	300	520
I_{GT} (mA)	65	65	100
I_L, max (mA)	150	150	150
I_H, max (mA)	100	100	100
dv/dt (V/µs)	1000	1000	1000
T_{GT} (µs)	2	2	2
T_j (°C)	−40 +125	−40 +125	−40 +140
Case type	TO-247AD	TO-247AD	TO-247AD

closed additional contacts shunting DPR's inputs, improvement of reliability can be achieved by connection of these contacts in series (Figure 5.20).

The capacitor of a R1C1 circuit and other elements of the device are improved and have a higher (fivefold) margin in terms of working voltage (Table 5.4).

As a voltage divider, three Zener diodes connected in series were selected with a stabilization voltage of 15 V and 10–20 W power each. With a rather low own energy consumption of the circuit, high-voltage margin prevents Zener diode's heating, improves reliability of their operation, and allows to improve absorption of high-energy overvoltage impulses. Parameters of the most suitable Zener diodes for these purposes are listed in Table 5.5.

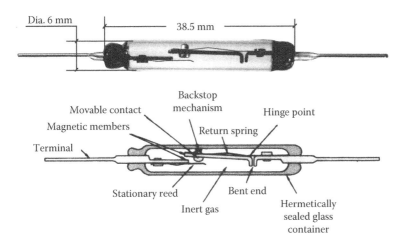

FIGURE 5.21
A powerful gas-filled RS Bestact R15U by Yaskawa Company with a two-stage switching.

TABLE 5.3

Main Parameters of Some Types of Changeover RSs

Parameter Type, Manufacturer	GC 1917 Comus	HSR-830R Hermetic Switch, Inc.	HSR-834 Hermetic Switch, Inc.	HSR-V933W Hermetic Switch, Inc.	DRR-DTH Hamlin
Max switching power (W)	60	25	100	100	50
Max switching voltage (V)	400	250	500	500	500
Max switching current (A)	1	1	3	3	0.5
Dielectric strength (V)	1000	1000	1000	1500	1200
Operate time (ms)	4.0	3.6	2.0	4.2	4.5
Release time (ms)	0.15	4.2	1.0	3.7	7.0
Balloon dimensions (mm)	$D = 5.6$ $L = 36$	$D = 5.3$ $L = 32$	$D = 5.3$ $L = 34$	$D = 5.3$ $L = 33$	$D = 5.5$ $L = 39.7$

TABLE 5.4

Main Parameters of Some Types of High-Quality Capacitors for RC Charging Circuit

Capacitor Type	Manufacturer	Capacity (µF), Voltage (V)	Dimensions (mm)	Min/Max Temperature (°C)
B43504B2477M	EPCOS	470, 250	Dia. 30 × 30	−40 +105
B43505A2477M	EPCOS	470, 250	Dia. 30 × 35	−40 +105
EETHC2E471CA	Panasonic	470, 250	Dia. 25 × 30	−40 +105
MAL215933471E3	Vishay	470, 250	Dia. 25 × 40	−25 +105
MCHPR250V477M25X41	Multicomp	470, 250	Dia. 25 × 41	−25 +105
381LQ471M250J022	Cornell Dubilier	470, 250	Dia. 25 × 30	−40 +105

TABLE 5.5

Main Parameters of Some Types of Power 15 V Zener Diodes

Parameter / Type	NTE5191A	1N2979	BZY93-C15
P_D (W)	10	10	20
V_Z (V)	15	15	15
I_{ZM} (mA)	560	560	1000
I_{ZT}	170	170	170
Z_{ZT} (Ω)	3	3	1.2
I_R (μA)	10	5	50
T_{OPR} (°C)	−65 +175	−65 +175	−55 +175
Case type	DO-4	DO-4	DO-4

The following types of devices can be recommended as VD4 trigger diode (Figure 5.3) with an opening (breakover) voltage of 24–36 V and current rate of 1–2 A: NTE6407, DB3, BR100/03, CT-32, HT-32, and others. As an electromagnetic relay Rel1 (Figure 5.2), we can recommend full-size crystal can relays with two changeover contacts (two normally closed contacts are used to increase reliability), which switch 2–5 A current with a coil for 24 V DC. As an example, the following relays can be mentioned: REN33, REN34, REK134, RES48, 782XDXH, H782, B07, FW, SF, G2A-434ADC24, HGPRM-B4C05ZC, 2B-7506, and others.

There is no need for fine-tuning of a pickup threshold of this device. It is important that its actuation is always earlier than that of DPR under any suspicious regime in a control circuit, since unwanted operation of the device as a result of incorrect setup does not influence operation of DPR protected by the device.

5.3.3 Reed Switch Relays with Adjustable Actuation Threshold

As mentioned earlier, the RSs are widely used components of equipment produced by several companies. The following advantages of RSs, such as air tightness, longevity, quick response, special gas medium or vacuum where contact points are located, lack of the necessity to adjust or clean the terminals, high level of galvanic separation between the input (control coil) and output (RS), and precise and consistent threshold of actuation (pickups), make these devices indispensable in certain automation systems and gauging equipment. Besides common and well-known types of use of the RSs, recently, a suggestion has been made to use them as sensitive elements of current and voltage in devices intended to decrease of vulnerability of DPRs to IRDIs.

The sensors of current and voltage based on RS relays installed in the aforementioned devices should be actuated in the event of emergency or near-emergency modes of operation of power equipment protected by DPR.

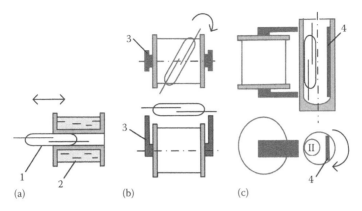

FIGURE 5.22
Design layout of RS relays with adjustable actuation threshold (pickup). (a) With in-line shift of an RS inside the coil, (b) with the shift of the RS's axis relative to the coil's axis and the outer placement of the RS, (c) with off-center shift of the RS and magnetic shunt. (1) RS, (2) coil, (3) ferromagnetic core, (4) ferromagnetic shielding plate (magnetic shunt).

Such a use of RS relays stipulates the need in the option to adjust the actuation threshold. However, none of the RS relays currently manufactured by the industry has this option, which has made it necessary to develop the design of this kind of RS relay.

Let's review the most appropriate designs (for this kind of use) of the RS relay with adjustable actuation thresholds; refer to Figure 5.22 [17].

The simplest variant is represented by the layout shown in Figure 5.22a with an in-line (axial) position of the RS and the coil and translational displacement of the RS along the coil's axis. The highest sensitivity level of the relay is achieved when the contact gap of the RS is located in the center of the coil. When this gap is shifted relative to the coil's center, the sensitivity of the RS to current is reduced. However, practical deployment of this design appears to be not that simple (see Figure 5.23).

In order to shift the RS, it was necessary to make a special element, similar to worm gear, where movement of threaded knob 1 around its axis triggers movement of a screw with an outer thread located in the end of plastic moving part 2 with a molded-in RS. Besides the complexity, another disadvantage of the design is a long length (*L*) of the relay, which exceeds the triple length of the RS's tube. Another disadvantage of this design is the fact that the RS is leaving the efficient magnetic shielding zone, when the RS is coming out of the coil. The strength of insulation between the coil and the RS in this design does not exceed 1 kV.

The design illustrated in Figure 5.22b is simpler from the production standpoint. There is a ferromagnetic core with poles inside the coil of a relay built according to this layout. The RS is resting on the outer side of the coil, and its axis is parallel to that of the coil.

The construction of the relay built according to this design is less complicated than the construction mentioned earlier (see Figure 5.24).

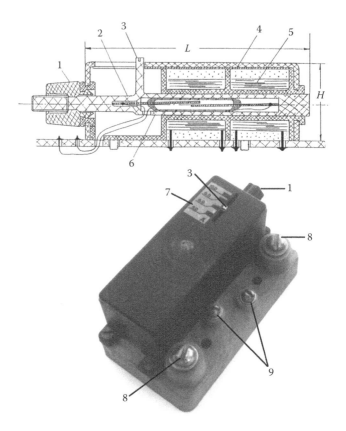

FIGURE 5.23

RS relay with adjustable actuation threshold with axial position of the RS and the coil and coaxial movement of the RS. (1) Moving control knob, (2) plastic moving part with molded-in RS, (3) the RS's position indicator, (4) ferromagnetic screen, (5) coil, (6) *RS, (7) scale, (8) winding terminals, (9) RS's terminals.

The RS position shown in Figure 5.24 ensures maximum sensitivity of the relay. Desensitization of the relay is achieved by turning the RS in such a way that there is an angle between the longitudinal axes of the RS and the coil. The minimum level of the relay's sensitivity is obtained at 90° between the aforementioned axes.

In order to adjust the actuation threshold in this design, it is necessary to turn the turning body 10 within the range of 0°–90° by means of the capsule's terminals and by the further fixing of this position by means of a securing nut 8. The relay can be mounted onto a surface by means of a nut 6 or by means of the extended flange with holes and ordinary screws. The relay of this type has a cylinder-shaped body of large diameter (exceeding the length of the RS's tube) and height, which is almost three times larger than the length of the RS. The insulation strength between the RS and the coil in such a relay is higher than that of the previous design and can reach as high

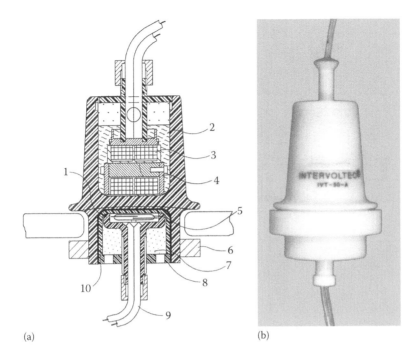

(a)   (b)

FIGURE 5.24
(a) RS relay with adjustable actuation threshold built according to (b) layout, where the RS is located outside the coil (the longitudinal axis of the RS is making an angle with a longitudinal axis of the coil). (1) Plastic mushroom-shaped capsule with a socket for an RS, (2) filling epoxy compound of the body's cavity with a coil, (3) coil, (4) ferromagnetic core, (5) RS, (6) nut, (7) filling epoxy compound of the turning body with an RS, (8) securing nut of the turning body, (9) RS's terminals, (10) turning body of an RS.

as tens of kilovolts. It is noteworthy that I manage to develop a design for voltages reaching as high as 70 kV (subject to the corresponding thickness of the insulation body, its length, and selection of corresponding insulation material for its production).

The most compact is the relay with adjustable actuation threshold (Figure 5.25), which is built according to the layout shown in Figure 5.22c. The RS and the magnetic shunt in this relay are located opposite to each other off-centered inside the turning tube. To achieve the maximum sensitivity, the RS should be as close to the poles of the control winding as possible, while the magnetic shunt should be as far away as possible. When the tube is turning, the RS is moving away from the core's poles, while the magnetic shunt is coming to its place reducing the magnetic flux in the RS's area. The use of this magnetic shunt enables obtaining a wide range of adjustments of the actuation threshold at small diameter of the turning tube, that is, to reduce the size of the relay. After the relay is set up for a specific actuation current, the position of the tube is locked by means of the securing screw 4. This relay also has high insulation strength between the RS and the control coil,

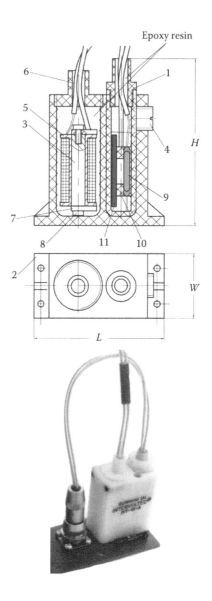

FIGURE 5.25

Design of a compact relay with adjustable actuation threshold with the off-center shift of the RS. (1) Extended part of the turning body tube, (2) fixing flanges, (3) ferromagnetic core, (4) securing screw, (5) coil, (6) coil's terminals, (7) core's poles, (8) pole mounting screws, (9) RS, (10) insulation spacers, (11) magnetic shunt.

especially when using wire in a high-voltage insulation to act as the coil's and RS's terminals.

Adjustable actuation threshold can be adopted in a differential relay as well, which responds to the difference of current and voltage ratings delivered to two different inputs of this relay (Figure 5.26).

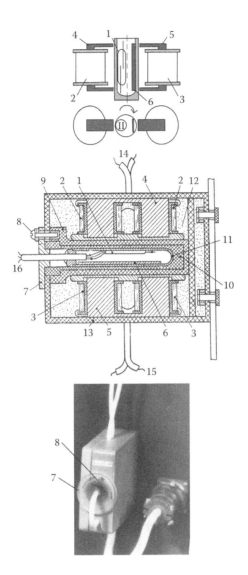

FIGURE 5.26
Differential RS with adjustable actuation current. (1) RS, (2, 3) coils (control winding), (4, 5) flat ferromagnetic Π-shaped cores, (6) magnetic shunt, (7) relay setup dial scale, (8) dial scale lock, (9) immobile insulator, (10) turning part of an insulator, (11) tube with the RS and magnetic shunt, (12) filling epoxy compound, (13) rectangular plastic body of the relay, (14, 15) control windings' terminals, (16) RS's terminals.

The design of this relay is actually an alternative of Figure 5.22c. However, the difference is that it has two coils located in one plane, but on opposite sides of the turning tube with the RS and magnetic shunt.

In this design when the dial scale is turning, the position of the RS 1 and magnetic shunt 6 is changing in relation to core's poles 4 and 5 of the control coils. When the dial scale is turning, the RS is moving away from one of the coils and approaches the other coil. As a result, the level of effect of these coils, that is, of the input signals, on the RS is changing. If the turn-on polarity of the coils is changed, the magnetic field intensity near the RS in the neutral position of the turning insulator 10 will approach to zero. When the insulator with the RS is turning, the effect of one coil on the RS will increase, while the effect of the other coil will decrease.

The design of this relay ensures high level of galvanic separation between inputs and the output due to the presence of the high-voltage insulator 9. If this insulator is molded together with the body from high-quality plastic and after the relay is assembled quality epoxy filling compound is used under vacuum, one can achieve insulation strength of tens of kilovolts in this design as well.

For operation with DPR, the insulation strength of tens of kilovolts is, of course, redundant. But the insulation strength of 5–10 kV of the impulse voltage will be beneficial when it is all about a device that protects from powerful electromagnetic impulses featuring high directed voltage. In the designs according to layout (Figure 5.22b and c), this level of insulation is easily adopted since these designs were initially developed to operate under high voltage ratings [6].

The aforementioned designs have been tested in practice and have shown good results both as overvoltage relay and as overcurrent relay. However, in some practical instances, when it is about their use in protection devices of digital relays, which operate as a remote protection, an undervoltage relay may be required.

In order to implement a function of undervoltage relay, there should be an additional winding L1 directly on the RS with relatively small number of windings. This winding will be connected to a stabilized power source of 5 V, while the operating winding L2 insulated from the RS will be connected via a diode rectifier VD2 and smoothing film capacitor C1 (see Figure 5.27), which possess very high margin of voltage (input voltage from a standard voltage transformer used in electric power industry usually does not exceed 100 V).

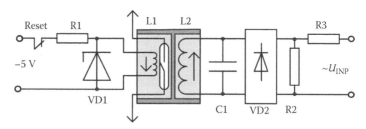

FIGURE 5.27
RS-based undervoltage relay.

The L1 and L2 windings are connected opposite to each other, and thus the total magnetic field in the area of the RS is close to zero under the normal mode of operation. In case of significant reduction of input voltage (to which distance protection of power transmission lines usually reacts in addition to current rate increase), the magnetic field of the L2 winding is reduced, while the magnetic field created by the L1 winding remains unchanged. The resulting magnetic field near the RS is increasing and it picks up. The design of the relay can be any of the aforementioned designs.

Diodes BY 2000 (Diotec Semiconductor) rated for 2000 V and 3 A (impulse current of 80 A) can be used in a device in a casing DO-201 (diameter 4.5 mm, length 7.5 mm). Capacitor MKP1T041007H00 (WIMA) 1 µF, 1600 V sized 24 × 45.5 × 41.5 mm. These elements are located on a printed circuit board of the device outside the relay's body. The use of these high-voltage elements at relatively low voltages entering the circuit after a splitter on the R2–R3 resistors (15–20 V) is necessary to ensure high resistance of the device to overvoltages, generated by a powerful electromagnetic impulse.

It is recommended using miniature vacuum RSs, which can sustain test voltages of not less than 1 kV and having their own actuation time about 1 ms in all the aforementioned relays (see Table 5.1).

It is recommended manufacturing of structural insulation elements of all the aforementioned types of relays from molded thermoplastic ULTEM-1000 (polyetherimide [PEI])—semitransparent material of amber color, which possesses several mechanical, temperature (−55°C + 170°C), and electrical (33 kV/mm, tgδ = 0.0012) features, low water absorption capacity (0.25% during 24 h), high resistance to different types of radiation, and a decent adhesion to epoxy compounds. Using STYCAST 2651-40 (Emerson & Cumming) as a filling epoxy compound is also recommended. This is a two-component black-colored compound, which possesses good dielectric features (18 kV/mm, tgδ = 0.02), low water absorption capacity (0.1% during 24 h), wide range of operating temperatures (−75°C + 175°C), very low viscosity in a liquid form, and good adhesion to metal and plastic surfaces. The linear expansion coefficient of this compound is close to ULTEM-1000, which is very important if the relay has to operate in a wide range of temperatures. CATALYST-11 should be used as a solidifying agent.

It should be noted that it is not allowed to fill the RS with an epoxy compound directly. Initially, it should be covered with a layer of a damping material, which will compensate mechanic stress emerging during curing of the epoxy compound.

The aforementioned technical solutions can be a practical foundation for the production of current and voltage relays with adjustable actuation threshold for devices intended to increase the resistance of DPRs to IRDIs as well as for other applications, where RSs with adjustable actuation threshold are needed.

5.3.4 Technical and Economic Aspects of Active Method DPR Protection

The proposed active method of DPR protection against IRDI is so unusual and so different from anything that has been known before, which experts inevitably have raised a barrage of questions and a tornado of emotions (alas, they are not always positive). Lack of the answers to many of the emerging questions in previously published articles often has led to misunderstandings and hence to complete rejection of the proposed method. Therefore, we will try to formulate the most frequently asked questions on this matter and answer them.

> Question 1. According to the diagram, the RSs are hung on DPR all around like fairy lights on the Christmas tree.

It is clear that the RSs are not hung on the inputs and outputs of DPR like "fairy lights," rather they are located inside the single-shielded enclosure together with all other elements of the proposed protection device, and the design of the enclosure is similar to that of the DPR with the only difference being that there is no need for a screen but there is an access to the operating threshold control units of the threshold element reed relays. This separate module has the same terminal blocks for connection to external circuits as the DPR.

> Question 2. There is a common perception that RSs are not reliable (they "stick"). Is it reasonable to use them in devices that should be characterized with enhanced reliability?

RSs, or rather reed relays used in the SU of protection devices, have a wide range of benefits compared to conventional electromagnetic relays. First, the contact points of dry RSs are enclosed in a sealed cylinder filled with a mixture of pressurized inert gases or vacuum, and so they are not exposed to adverse environmental factors (moisture, dust, gases). These contacts require no adjustment or cleaning during the whole lifetime. Second, the reed relay is three to five times, or more, faster than the conventional electromechanical relays. Third, under AC, such a relay has a reset ratio of 0.9–0.95, which is much higher than for conventional relays. Fourth, the reed relay allows for easy approach to the galvanic isolation level between input and output (between coils and contacts) of tens of kilovolts, which is unattainable for conventional electromechanical relays [12]. Fifth, unlike the conventional relays, the reed relays have clear and stable pickup thresholds under gradual increase of control coil current/voltage, thereby enabling the development of sensitive reed-based measuring units for protection purposes. In addition to the aforementioned, it should be noted that the dry RSs are insensitive to the position in space and can be easily combined with electronic, electromagnetic, and magnetic components to develop a number of different functional modules and devices on their basis [18].

High-quality vacuum and gas-filled RSs manufactured by leading companies specializing in this industry (such as those earlier proposed for use) are not cheap ($15–$30 each), but they are highly reliable and widely used not only in industry and communications but also in military and aerospace. Thanks to a number of advantageous options, the RSs occupy an intermediate position between semiconductors and electromechanical switching elements. Therefore, the automatic telephone exchange (ATX) such as reed-based ATX ("Quant"), are called "quasi-electronic." According to the specifications, the lifetime of such ATX is 40 years and the number of failed reeds should not exceed 0.3% within the period. So the figures speak for themselves.

However, the reed relays have one fundamental difference from the conventional electromechanical relays: their magnetic system is not isolated from the contacts; rather, it is formed by them. This difference causes low overcurrent capacity of the RSs. Unlike the conventional relays, the RSs do not withstand even the short-term contact overcurrent. This is due to fact that the magnetic field of the current flowing through the closed contacts of the RS is directed opposite to that of the coil magnetic field, holding the contacts closed, and weakens it, thus decreasing the contact force up to the appearance of the gap. This leads to the increased erosion and sometimes to welding of the reed contacts even under the short-term currents exceeding the maximum allowable value for particular type of the unit. Poor awareness of this aspect and ignorance of the reed differences from the conventional relays (with regard to the overload capacity) often lead to the equipment failures and, as a consequence, to the distrust of the RSs. Under properly selected operating mode, the RSs provide reliable circuit switching throughout millions of operation cycles.

When using RSs for switching external circuits with a wide variation of the current, no one wants to monitor the reed current operating mode. It is much easier not to use them at all, which often happens in practice. In the proposed construction, some RSs are included only in the device's internal circuits, where the current load is 10-fold less than the maximum allowable reed current. Other RSs turn off the digital input circuits with the maximum current of several milliamps, which is two orders less than the limit. And the current of several amperes can flow only though the RS connected in series with the output terminals of the DPR intended to switch on the CB trip coil. However, (1) these RSs do not switch these currents directly (they only assemble the circuit under no-current condition), and (2) their type (Bestact R15U, produced by the Japanese company Yaskawa) provides high current margin.

Question 3. Modern DPRs combine 10–20 and more different functions in a single module. Does it mean that the proposed protection device must contain the same number of input relays?

No, it does not. The point is that the variety of DPR functions embedded in a single terminal is based on the measurements of current, voltage, and angle between them. Accordingly, the input relays of the proposed protection device must contain threshold elements for current, voltage, and angle between them. Thresholds of pickup of all these elements must be less than the minimum values selected as the DPR set points.

> Question 4. Why do we need to use expensive DPRs together with some new and also expensive protection devices, if we can just go back to the cheap and resistant to IRDI electromechanical relays?

Indeed, the electromechanical protection relays (EMRs) have been operated for over a hundred years and still provide reliable protection against emergency operation for all types of electrical equipment. Suffice it to say that some large and diversified national power systems (e.g., in Russian) are even today nearly 70%–90% equipped with EMRs. However, despite the fact that EMRs have proven their high reliability, about 30–40 years ago, all the world's leading manufacturers of protective relays stopped developing and improving EMRs and began to intensively develop first the solid-state relays completely duplicating the functions and characteristics of EMRs and then the microprocessor-based relays with advanced features and improved performance. About 20–25 years ago, most of the world's leading manufacturers of protective relays stopped producing EMRs and concentrated all their efforts on the DPRs. The main reason for this phenomenon was that it was much more profitable to produce and test printed circuit boards with electronic components on the automatic equipment than to produce miniature mechanical elements on high-precision turning and milling machines and manually assemble them into the rather complex mechanical design and make manual tests and customization.

Due to the large difference in production costs between the EMR and the DPR, the consumer stands to gain too as today the cost of DPRs produced by world's leading relay producers is much less than the cost of EMRs with similar characteristics. The statement that today the EMRs are much cheaper than DPRs is not correct and is not supported by the analysis of the world market prices. For example, if electromechanical relay for three-stage line distance protection type LZ31 (made by ABB) could cost about USD 30,000–35,000 (according to current prices), its microprocessor-based analog with improved characteristics, such as the relay type D30 (made by General Electric), costs only USD 7,500, and the Chinese analog of the relay (type GTL-823 made by Guatong Electric) costs even less—only USD 5,000 today.

In addition, a powerful advertising campaign pursued by the developers of DPRs, universities, and research organizations interested in financing of new projects did the trick. Today, to raise the question about returning to EPRs means to become an outcast in the community of experts and to gain a character of retrograde who is trying to stop technological progress. None of the experts or decision-making officials would take such a responsibility. And even if they take it, it is safe to say that they will be inundated with charges of obscurantism and incompetence. In addition, for the sake of objectivity, it should be noted that DPRs do have some features and functionalities unattainable by EMRs.

With all of these factors in mind, it can be stated that the question of returning to the EMR is not relevant.

> Question 5. Suppose that the return to EMRs is really not possible today. But why not use DPRs completed with these EMRs instead of inventing some new RS-based devices?

In fact, the combination of EMR and DPR has long been used in practice (Figure 5.17).

However, they are not connected in series (as suggested)—they are connected in parallel to duplicate each other in order to increase the reliability. As explained previously [5], such a method of DPR and EMR combination (i.e., in parallel) is essentially not correct. In parallel connections, the EMR really must completely duplicate the DPR functions and have the same set points. In any combination of multifunction DPRs and EPRs, the whole set of expensive EMRs should be used, thus making this project very doubtful because of its high cost and availability of large areas for installation of a large number of different EMRs. The suggested protective reed-based device should be much more simple, smaller, and cheaper than a set of EMRs necessary to protect one DPR. And this is the only way to make its use promising.

> Question 6. To provide the versatility and full functionality, the functional capabilities of the suggested protective device should be the same as of the set of EMRs. Hence, its cost should be about the same. Why will it be cheaper?

Let's look at how the EMR works. For example, let's consider current-dependent time delay relay—inverse definite minimum time (IDMT). It is an electromechanical induction disk-type relay where an aluminum disk begins to turn slowly and the movable contact associated with this disk starts to approach the fixed contact under the certain threshold current. After some time determined by the speed of disk rotation (based on the current flowing through the relay coil), the contact closes (via the intermediate relay) the circuit of the CB trip coil.

The starting element of the proposed DPR protection device requires no current-dependent time delay. This starting element should trip only under the certain current, somewhat smaller than the disk pickup current. That's all. Other functions are not required, since all other functions will be done by the activated DPR. That is, in this case, instead of a complex and expensive IDMT relay, we need only the simplest relay consisting of a coil and an RS.

As another example, let's consider several types of line distance protection relays. The electromechanical alternative of this relay (e.g., type

FIGURE 5.28
Line distance protection relay LZ31 type.

LZ31 [Figure 5.28]) contains many complex and interconnected electro-mechanical assemblies providing three or four stages of line impedance measurement to the short-circuit point corresponding to these delay stages, special form of characteristics, etc. As noted earlier, the cost of such relays is USD 30,000–35,000. However, the whole complex is actuated by the simplest starting element controlling the balance between line current and voltage (Figure 5.29). The element is actuated by an imbalance between preset current and voltage values.

Large and complex distance protection relay types RYZKB, RYZOE, and RYZFB, manufactured by Allmanna Svenska Elektriska Aktiebolaget (ASEA) in the 1970s (Figure 5.30), implement several protection features. However, all these relays are equipped with very simple starting element (see the diagram in Figure 5.30).

These starting elements were an integral part of complex structures and were not sold separately. The exceptions were some types of relays, which were produced in Russia, for example, the relay type KPC-112 (Figure 5.31) containing special chokes and four-pole inductors with rotor. In essence, this relay is a separate starting element of distance protection. However, it is too complex, expensive, and large. Anyway, the combination of such obsolete designs with the up-to-date DPR technology is hardly a good idea.

In this respect, starting element of distance protection of type HZM (Westinghouse) could be much more appropriate (Figure 5.32).

This is a very simple device comprising T-shaped core with swinging rocker (upper part of the letter "T") and two coils: voltage coil and cur-rent coil acting on the ends of the rocker. The position of the rocker with

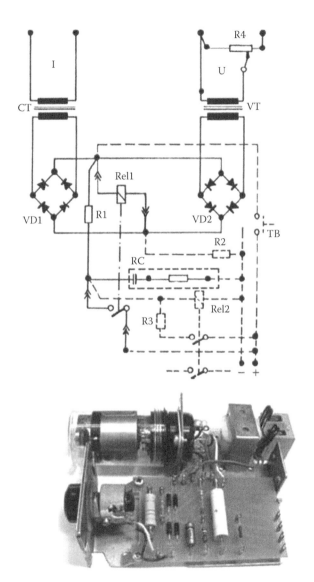

FIGURE 5.29
Principle of operation and design of the starting element of the distance protection relay LZ31.

attached contact depends on the balance of the magnetic fields generated by the current and voltage coils. This assembly is an internal part of HZM relay design and has never been sold separately.

The reed relay, built on the same principle of balance between current and voltage (Figure 5.33), is much simpler and more reliable. This relay responds to the difference between the magnetic fields generated by the current and voltage coils, and its threshold can be adjusted within a wide range by turning the reed capsule. Such a starting element can be successfully used in the SU.

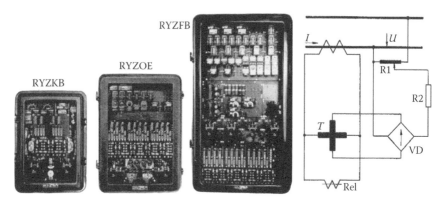

FIGURE 5.30
Line distance protection electromechanical relays of various types made by ASEA and the diagram of their starting element (produced in the 1970s).

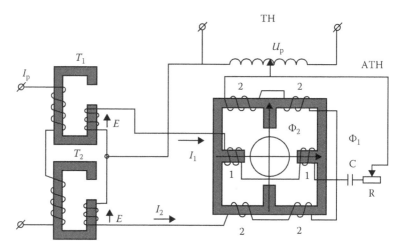

FIGURE 5.31
Relay KPC-112 (ChEAZ, Russia) with the induction mechanism.

Thus, the proposed device with a small number of reed-based elements of current, voltage, and the difference between them is much simpler and cheaper than a full-featured set of EMRs. In addition, reed-based starting elements do not require maintenance during operation, have significantly less delay within the overall relay response time, and provide a high level of insulation between input and output unattainable for old EMRs.

Question 7. In some cases, the CB trip command is issued directly by the protective relays (such as transformer gas protection relay) and is simultaneously duplicated by the signals sent to the logic inputs of DPR, thus triggering the fault recorder. How then will the proposed device (which blocks logic inputs of DPR) work?

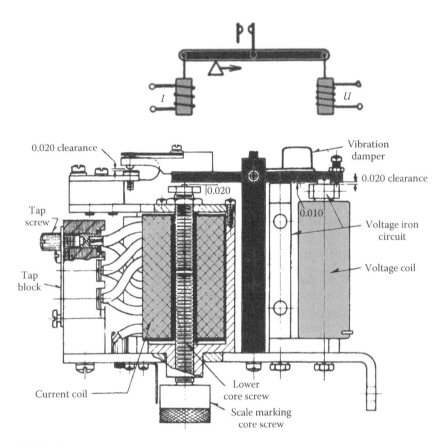

FIGURE 5.32
Balance electromagnetic starting element used in distance protection relay of type HZM (Westinghouse).

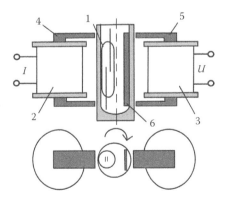

FIGURE 5.33
The simplest starting element of distance protection with adjustable threshold. (1) RS, (2, 3) coils with control windings, (4, 5) U-shaped flat ferromagnetic cores, (6) magnetic shunt.

This is easy to resolve: it is only necessary to send a signal from trigger relay contacts (in this case, gas relay) also to one of the inputs of SU of protection device. In this case, the DPR is unblocked and the fault recorder starts operation and records gas relay trip information.

Question 8. There is a requirement of inadmissibility to include any additional locking elements into the circuit of breaker trip coil, and in the proposed device, this circuit is switched by the contact of the addition auxiliary relay. Is this acceptable?

In fact, normally the open contact of an auxiliary relay is not connected into the trip coil CB. It is connected into the circuit connecting the output DPR relay contact to the trip coil of the CB. That is, this additional contact does not block the control circuit of the CB trip coil; it only blocks the output circuit of DPR. The control circuit of the breaker trip coil remains free to connect any external contacts or manually operated keys.

Question 9. How should one deal with complex protection units, for example, with the protection units providing the offset from the transformer excitation current inrush and containing filters of 2 and 5 harmonics? Should the proposed device also contain such filters? Or another example is the differential protection. How should one ensure the device operation if the emergency mode exists only in the protected area?

No, the SU does not need such filters or excitation current inrush offset for operation. The tripping of the SU under the transformer excitation current inrush only unblocks the DPR for about 10 s and nothing else. The DPR will block against unnecessary pickups with its own algorithm. After 10 s, the SU reverts to the original state and blocks the DPR again. The same applies to the differential protection. The SU device does not care where the fault is, within or outside the protected area. The only important thing is the presence of short-circuit current, while the damage zone will be determined by the DPR after the SU unlocks it. The SU response time is about 6 ms and almost does not affect the total response time of RP since the DPR proper time is 30–40 ms.

Question 10. If the DPR and the EPR are connected in series, the capabilities of RP will actually be limited by the capabilities of the EMR, as it has reduced capabilities and characteristics. Is this good?

No, it is not. The proposed device in no way defines any properties or characteristics of RP. It only unblocks the DPR at the moment when at least one parameter of the entire set of monitored parameters approaches the DPR set point. The subsequent behavior of the RP and its response to emergency mode will be completely determined by the properties and characteristics of the DPR.

It is obvious that in practice there are more complex modes of DPR operation, which are not discussed in the article, and such modes will require special starting elements to be developed. This is possible. However, even if this will require the development of such a special starting element, then based on a combination of RSs and magnetic circuits, it is possible to create simpler, cheaper, and faster units compared to the traditional electromechanical relays. For example, the device shown in Figure 5.33 may be well used to control the angle between the current and the voltage or as power measuring element.

The combination of magnetic, RS-based, and high-voltage discrete semiconductor elements provides additional opportunities. For example, Figure 5.34a illustrates the simplest device responding to the current difference, and Figure 5.34b illustrates the device with differential current desensitization by the DC value (offset).

Thus, the preceding analysis clearly shows that the practical implementation of the proposed method of DPR protection is quite feasible from the technical and economic point of view. Certainly, such an implementation should be done by the manufacturers of DPRs, which can offer consumers a quasi-electronic SU as an option for improving the safety and reliability of the RP for critical objects.

> Question 11. High-frequency and short impulse distortions can also enter DPR through current and voltage circuits. How can we protect them using the proposed method?

The issue about protection of current and voltage circuits and specific technical solutions, which ensure this protection, requires additional research. The fact is that input current and voltage enter the electronic circuit of DPR through its own imbedded input current and voltage transformers, which are available in each DPR and which transform rather powerful input rates into low-powered signals. These signals are on the order of a very few volts that enter the analog-to-digital converter (ADC), where analog signals are quantized based on their level and converted to a digital code. Quantization

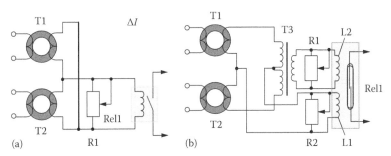

FIGURE 5.34
Variants of quasi-electronic starting elements of differential protection: (a) with common and (b) with separate balancing elements (potentiometers).

(sampling) takes place in DPR at a relatively low frequency of 600–1200 Hz. Thus, the process takes some time, which is several orders longer than the duration of DPR's impulses. During the IDEI impulse, the ADC will not be able to make necessary conversions. This is why the nature of IDEI impulses impact on DPR, which entered through current and voltage circuits, will be determined by inductive and conductive interference, which almost do not differ from the same interferences entering the DPR by other means. However, it is possible to weaken these interferences by the placement of grounded foil between the primary and secondary windings or disconnecting and bypassing the secondary current and voltage circuits of the aforementioned internal input transformers. In order to do this, the DPR manufacturer has to take these circuits out to a separate external connector so that RSs of the proposed protection device for the circuits can be connected through them.

5.3.5 Power Transformer Protection

The power transformers differ from the electronic devices, also exposed to damage under such impacts, in their impossibility of quick change in case of failure (see Chapter 3). In the context of the foregoing, it becomes clear that it is important to protect power transformers against damage under geomagnetically induced currents of low frequency.

One of the solutions proposed by Western scientists is to include a current-limiting capacitance (i.e., powerful capacitor) in the transformer's neutral. Elements of this type transmit AC of the mains' frequency (generated by asymmetry of the phase current) under normal operation but block the flow of the low-frequency geomagnetically induced currents. Some Western sources even give the cost of such current-limiting elements (it is about USD 40,000) and comment that presently the congress is considering the question of investing USD 150 million in the installation of such elements on the most important power transformers. The major problem of this solution is the generation of the extremely high voltages on such elements under short circuits in the mains (we are speaking of power transformers of 110 kV and higher).

To solve the problem, Ref. [19] proposes using special super high-power cold cathode electron tubes, short-circuiting the capacitor undervoltage buildup and allowing the short-circuit currents of dozens of kA (see Figure 5.35).

The installation of a powerful capacitor with super high-power electron tube (type 4275 Bi-Tron™, 30 kV, up to 75 kA) also requires using additional switching devices S1, S2, and S3, needed to connect the unit to or disconnect it from the neutral of the power transformer. Generally, it appears to be a quite massive and very expensive device that will unlikely be widely used due to its high cost.

We, on the other hand, propose another type of the protection of high-power transformers against the low-frequency geomagnetically induced

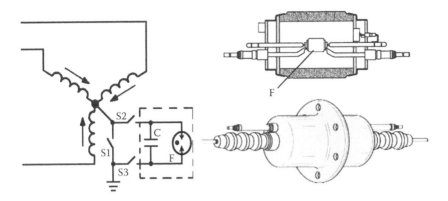

FIGURE 5.35
Protection of the power transformer against the low-frequency induced currents based on the power elements included in the neutral, proposed by Western specialists. (Adapted from Kappenman, J.G. et al., GIC mitigation: A neutral blocking/bypass device to prevent the flow of GIC in power systems, IEEE PES Special Publication 90TH0357-4-PWR, Special Panel Session July 17, 1990, pp. 45–52.)

currents using a special relay containing no microelectronic components and based on discrete high-voltage elements [17] resistant to electromagnetic interferences and surge overvoltage (see Figure 5.36).

Figure 5.36a shows the operational principle of the relay sensitive to the DC component in the power transformer neutral and insensitive to the widely varying AC component.

The relay consists of an RS with a coil placed on the cable (bus) that connects the transformer neutral to the grounding point perpendicular to the axis of the cable and a conventional toroidal current transformer (CT) installed on the same cable.

If there is no DC in the neutral the magnetic field of the cable (bus) acting directly on the RS, this is fully compensated by the magnetic field of the coil put on the RS and powered by the CT. AC changes in the neutral lead proportionally to the changes in both magnetic fields acting on the RS and to their compensation. Under high DC in the neutral (over 10–20 A), the balance of the magnetic fields acting on the RS is offset: the magnetic field of the cable (bus) still acts, while the compensating magnetic field of the coil energized by the CT is disabled as the DC component of the current is not transformed by the CT. This leads to RS activation. The real relay circuit includes an additional output amplifier installed on VS thyristor, varistor R_U, and the R1C1 all protecting the thyristor from interferences and voltage surges (see Figure 5.36b). The relay is equipped with a continuous electrostatic shield and a ferromagnetic shield with the only window on the cable side next to the RS and is connected to the circuit of the CB switch trip coil through a special twisted-pair control cable with the combined multilayer shielding grounded at both ends and resistant to the electromagnetic pulses. The relay can be constructed on miniature

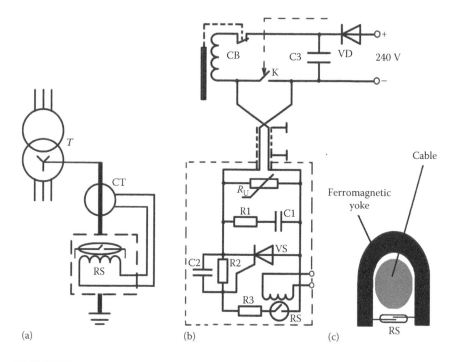

FIGURE 5.36
Relay protecting the power transformer against the low-frequency geomagnetically induced currents in the neutral circuit. (a) Principal diagram, (b) detailed circuit diagram, and (c) fixing of the reed switch (RS) on the neutral circuit cable.

high-voltage vacuum RSs, for example, of type KSK-1A85 (manufactured by Meder Electronics), with the electric strength of insulation between the contacts of 4000 V and the bulb having a diameter of 2.75 mm and length of 21 mm. This RS is capable of switching loads up to 100 W (the maximum switching voltage is 1000 V; the maximum switching current is 1 A) with the response time of 1 ms and a maximum sensitivity of 20 A. Additional ferromagnetic elements (magnetic field concentrators) located next to the RS can be used to increase the sensitivity. To get a relay with lower sensitivity and a higher pickup, the longitudinal axis of the RS should form a nonperpendicular angle to the axis of the cable on which it is installed.

The thyristor should also be miniature and of high voltage, for example, of type SKT50/18E (manufactured by Semicron), with a maximum voltage of 1800 V and maximum continuous current of 75 A, and must withstand high rates of voltage rise (1000 V/μs) under a wide operating temperature range (−40°C … +130°C). The power circuit of the trip coil is equipped with storage capacitor C3 enabling switch activation even under the loss of operating voltage. The R2C2 in series is designed to further enhance the immunity of the device. Capacitor C2 provides a certain delay of the thyristor switch on, preventing it from unlocking under the powerful impulse noise.

Application of the discrete high-voltage components instead of conventional microelectronics in the relay ensures its high reliability under powerful electromagnetic interferences and surge voltages specific to solar storms and electromagnetic pulses.

The purpose of this article was to identify the problem and present a simple solution. For this end, we have described a relay design to demonstrate its general concept only. It is obvious that the described device can be supplemented by signal relay (blinker) registering the response, time delay device, RS sensitivity control unit, reinforced insulation [12], etc. We believe that the proposed solution is more than adequate and far cheaper to implement than other proposed solutions.

5.3.6 Increasing Security of Remote Control of Circuit Breakers from Intentional Destructive Impacts

Due to a negative trend of adding extra functions to DPRs, which have nothing to do with RP functions (this issue was touched upon earlier [20]), implementation of the suggested protection principles of DPR will be complicated in some instances. I am referring to the wide usage of DPR for RCCBs [21–23]. Obviously, the usage of the DPR in this way has nothing to do with functions of RP, but rather remote control of DPR via communication channels in order to change the CBs position is difficult to distinguish from a cyber attack by means of hardware.

As mentioned earlier, the problem of increasing the reliability of RP cannot be solved when DPR functions are combined with those that have nothing to do with RP, such as power equipment online monitoring and RCCBs. DPR should be used to fulfill objectives of RP only. A fortiori, in order to solve other problems, such as monitoring of power equipments, many specialized devices (from simple relays, for supervision continuity of trip coils' circuit of a circuit breaker, to sophisticated complexes, which provide real-time control of the composition of gases dissolved in transformers' oil or the level of partial discharges in insulation) are available in the market. I think that RCCB should also be separated from RP and be performed by separate hardware. This is the only way we can increase the reliability of RP and ensure its efficient protection from IRDIs. This separation will not only allow providing highly efficient protection of DPR but also implement a protected remote control of circuit breakers (PRCCBs).

The suggested PRCCB system (see Figure 5.37) is a hybrid and combines both a digital controller with a network channel of data transfer and a cable channel with an electromechanical relay. The main purpose of the system is to prevent an unauthorized change of CB position during a cyber attack and failures of electronic devices incorporated in the system. The secondary objective of the system is to increase its survivability and maintain its working capacity after IDEI. The overall idea of the system is that any command for changing the CB position is transferred via a computer network

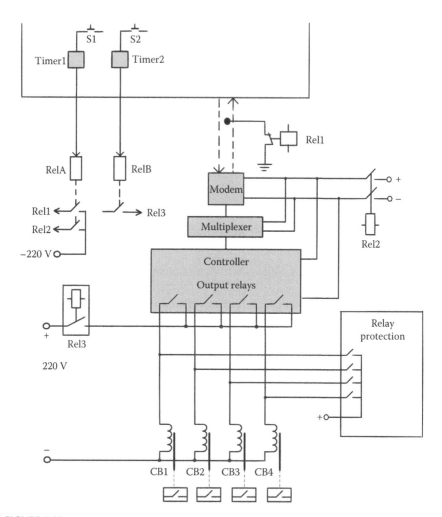

FIGURE 5.37
The offered structure of protected remote control system of CB. Power supply circuits and CB trip coils are shown provisory to simplify the diagram.

and should be confirmed by short pickups of an electromechanical relay at a substation by energizing its coil via an ordinary control cable. Why is it necessary to use the electromechanical relay, and why can't we use a communication channel based on fiber-optic communication system (FOCS) for the confirmation command?

The problem is that FOCS does not provide protection from IDEI, since they are equipped with complex digital multiplexor switches on both sides, which translate electric signals into light signals on one end of FOCS and restoration of electric signals on the other end. Our research of several types of multiplexor switches showed [24] that they cannot bear even a standard high-voltage impulse as required by electromagnetic compatibility standards.

If the internal microelectronic components of the FOCS are broken under IDEI, the condition of their output circuits will be unpredictable. If a broken FOCS will not be able to send a remote command for the CB, there will be no disaster, but if its output circuits were actuated, the problems will be inevitable. The same is applicable to all other components of a PRCCB system (modem, controller, etc.).

Besides, since fiber-optic communication channels with all mating equipment are rather expensive, there is a tendency to refuse the use of dedicated FOCS and use existing computer network channels based on twisted-pair wires instead. Moreover, in order to additionally reduce costs of control systems, RP, and automatics even more, a transition to Wi-Fi technology is under thorough investigation now. In any event, many of the world's leading manufacturers of DPR already produce them with built-in Wi-Fi modems. In fact, the idea of converting all power electrical equipment to communicate over standard computer networks, including wireless, is a central idea of the smart grid concept. In connection with this, the development of special hardware (which is not connected with computer networks and is highly resistant to IDEI) for increasing security of RP and CB remote control system from IRDIs, including electromagnetic impacts and cyber attacks [25], has become relevant. This is why electromechanical relay controlled by auxiliary voltage via control cables to act as elements of such protection is selected. In order to protect the additional communication channel from malicious external connection and unauthorized actuation of electromechanical relays, the wires of different control cables are used in addition to two relays: RelA and RelB instead of one (see Figure 5.37). Of course, the coils of these relays and current leading cable wires through which the relays are powered should be protected (e.g., by means of variable voltage resistors) from high-voltage impulse events, which can be applied to these wires under powerful electromagnetic impulse of IDEI. In addition, powering these relays by AC of industrial frequency with a capacitor connected in series and a splitting transformer used on the side of a substation is recommended. These measures will allow preventing actuation of RelA and RelB from current of extremely low frequency applied to underground cables under electromagnetic impulse's E3 component's effect [25].

In the suggested system, any command for changing the CB position transferred via a network channel of any type should be accompanied by a short-term, remote turn-on of RelA and RelB via the control cable. The contacts of these relays energize local electromechanical auxiliary relays: Rel1 (unblocks the network communication channel), Rel2 (supplies power to electronic devices of the system), and Rel3 (turns on the power supply circuit of CB trip coils). These local relays can be different in terms of their characteristics. For example, Rel1 is a high-frequency relay, and Rel3 is a relay with powerful contacts designed for switching the inductive load on DC. The availability of two control relays—RelA and RelB—with separate control channels increases the protection of the system from unauthorized access.

The first to pick up is RelA; after the necessary information about the changing position of a certain CB that is transferred to the controller and the closing of contacts of a corresponding output controller's relay, RelB will be pickups and energize Rel3 by its contacts. The time during which RelA and RelB remain energized is controlled by timers in order to prevent continuous engagement of these relays because of human error. In fact, this is a short period of time during which it is almost impossible to undertake an effective cyber attack. The blocking of the communication channel and disconnection of the controller's power supply beyond this short time period eliminates the danger of preliminary actuation of output controller's relays as a result of a cyber attack with the further unauthorized changing of CB position when electromagnetic relay RelB is energized. The same measures will dramatically reduce the probability of failures of sensitive electronic equipment (modem, multiplex, and controller) as a result of IDEI.

After a registered cyber attack or electromagnetic impulse, the remote control of the CBs should be prohibited until a special check since the condition of the controller after such impacts is not known.

The output controller's relays can be standard and low voltage; this type is usually installed on controllers. However, Rel3 should have contacts that can be compatible with switching powerful inductive load (CB trip coils) at DC 230 V.

Analysis of specifications of widely spread electromagnetic relays shows that the majority of them are not designed for the switching (and even turn-on) of inductive loads at 230 V DC [26]. Specially designed relays are used for this purpose: they ensure multiple series breaks in a switched circuit (Figure 5.38), or they accommodate a continuous magnet near the contacts,

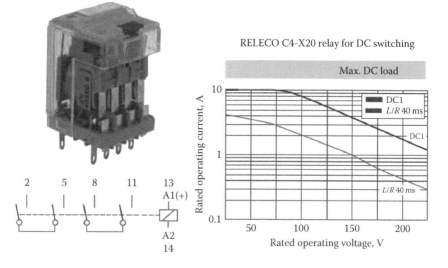

FIGURE 5.38
The C4-X20 (RELECO) relay type (with partially removed casing) with two double break contacts and high switching ability at DC.

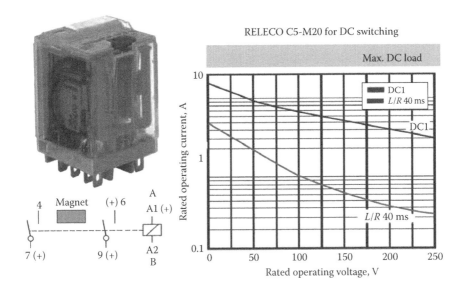

FIGURE 5.39
The C5-M20 (RELECO) relay type with two make contacts and a blowout magnet for increasing switching ability for inductive load.

designed to push out the electric arc from the intercontact gap (Figure 5.39). There are also relays with triple series breaks per contact (Figure 5.40), which can control trip coils of old-style high-voltage CBs consuming high-ampere current.

In those cases where the use of a control cable to control RelA and RelB is not possible due to the remoteness of a control station from a substation, it is possible to use FOCS as a permitting communication channel. However, the decreasing of the system's protection level from IDEI should be kept in mind. In order to prevent the occurrence of spontaneous commands for changing CB position due to the failure of electronic equipment of the system, they should be equipped with a self-diagnostic feature: the FOCS channel should be equipped with a continuous monitoring of its own soundness as well as the position of RelA and RelB, whereas the controller should be equipped with a self-diagnostic system, which is automatically actuated immediately after Rel1 and Rel2 actuation including a status scan of output relays (they should be off) and the validity of the operation of the communication channel.

When a failure is detected, the self-diagnostic system should block further operation of the controller. The use of the system for RCCBs should only be allowed when the control station receives information about operating efficiency of all the elements of the system.

The suggested system will dramatically increase security of RCCBs from IDEIs and cyber attacks. You can see that in both cases, that is, for DPR

FIGURE 5.40
The RMEA-FT-1 (RELEQUICK S. A.) relay type with one triple make contact, capable of switching currents up to 3 A in an inductive load at 220 V DC.

protection and for RCCB protection, electromechanical relays are used. However, the use of these relays is different due to the difference in the operating algorithms of DPR and PRCCB. In the first case, the command to the CBs is sent automatically when the controlled operation mode of power network or utilities equipment is changed, whereas in the latter case, the command is sent manually by operators working at the control station. This results in different principles of implementation of protection. In the first case, it is important to protect the DPR continuously working in the automatic mode from unauthorized changes of its settings or internal logic, which induces actuation of output relays, and it is not possible to check the validity of commands before actuation of the relays. Besides, it is not possible to send a sort of a permission signal to the DPR in case of an emergency mode in the control circuit. This permission signal should be generated on the spot as a result of the power network emergency mode. However, in the latter case when the protected object (PRCCB) is not operating in automatic mode, the task becomes much easier making it possible to use an external permission signal. In addition, in critical situations, the PRCCB can be cancelled completely. These natural differences in principles of implementation of protection from intentional destructive impacts justify splitting of tasks of RP and RCCBs.

References

1. Gurevich V. I. Electromagnetic impact on microprocessor protective relays, Part 2. *Components and Technologies*, 3, 2010, 91–96.
2. Baydyn F. N., Nikitina V. N., Safronov N. B. Electrophysical parameters and radioshielding properties of magnesian-shungite composite construction materials. Reports of *Ninth Russian Scientific and Technical Conference on EMC of Technical Means and EM Safety*, St. Petersburg, FL, 2006, pp. 292–294.
3. Gurevich V. I. The new concept of microprocessor protective relays design. *Components and Technologies*, 6, 2010, 12–15.
4. Schedrikov B. D. Eletromechanical relay protection devices in power industry: Present and future. *Relay Protection and Automation*, 1, 2010, 61–63.
5. Response of V. I. Gurevich to opponent protection engineers. *News in Power Industry*, 1, 2009, 41–42.
6. Gurevich V. I. About of a some ways for solving problem of electromagnetic compatibility of protective relaying. *Industrial Power Engineering*, 3, 1996, 25–27.
7. Gurevich V. I. Principles of improving immunity of static current relay. *Power Industry and Electrification*, 2, 1992, 16–18.
8. Gurevich V. I. Improving EMC of relay protection in power industry. *Industrial Power Systems*, 2, 1995, 48–50.
9. Gurevich V. I. About development of relay protection in electric mains. *Power Building*, 1, 1994, 48–51.
10. Gurevich V. I. New generation of universal protection relay of maximal current. *Electrotechnics*, 1, 1994, 61–66.
11. Gurevich V. I. Hybrid reed-semiconducting devices—New generation of protection relay. *Challenges of Power Industry*, 9–10, 2007, 27–36.
12. Gurevich V. *Protection Devices and Systems for High-Voltage Applications.* Marcel Dekker Inc., New York, 2003, p. 292.
13. Gurevich V. I. High-stable fast-speed reed-semiconducting current relay. *Energo-Info*, 2, 2007, 84–88.
14. Gurevich V. I. Electromechanical and digital protective relays. Whether the symbiosis is possible? *Relay Protection and Automation*, 2, 2013, 75–77.
15. Gurevich V. I. Device for protection of relay protection. *Scientific Journal of Electrical Engineering*, 3 (3), June 2013, 52–57.
16. Gurevich V. I. Increasing security of remote control of circuit breakers from intentional destructive impacts. *Power Transmission and Distribution*, 5, 2013, 114–117.
17. Gurevich V. I. Reed relays with adjusted pickups. *Components and Technologies*, 11, 2013, 30–33.
18. Gurevich V. *Electronic Devices on Discrete Components for Industrial and Power Engineering.* CRC Press, Boca Raton, FL, 2008, p. 420.
19. Kappenman J. G., Norr S. R., Sweezy G. A., Carlson D. L., Albertson V. D., Harder J. E., Damsky B. L. GIC mitigation: A neutral blocking/bypass device to prevent the flow of GIC in power systems, IEEE PES Special Publication 90TH0357-4-PWR, Special Panel Session July 17, 1990, pp. 45–52.

20. Gurevich V. I. Technological advantages in relay protection: Dangerous tendencies. *Electrical Engineering & Electromechanics*, 2, 2012, 33–37.
21. *MRI(K)3-C—Digital Time Overcurrent Relay with Control Function and Auto Reclosing*. Woodward SEG GmbH & Co. KG, Kempen, Germany, p. 77.
22. Circuit breaker controller—with arc flash mitigation. *PAC World*, Autumn 2007, p. 79. http://www.pacw.org/issue/autumn_2007_issue/news/industry_news/circuit_breaker_controller_with_arc_flash_mitigation.html.
23. Relion® protection and control. Generation, transmission and sub-transmission. ABB AB Substation Automation Products, Västerås, Sweden, 2013.
24. Gurevich V. I. Actual problems of the relay protection: Alternative view. *Electric Power's News*, 3, 2010, 30–43.
25. Gurevich V. I. *Digital Protective Relays: Problems and Solutions*. Taylor & Francis Group, Boca Raton, FL, 2010, p. 404.
26. Gurevich V. Peculiarities of the relays intended for operating trip coils of the high-voltage circuit breakers. *Serbian Journal of Electrical Engineering*, 4 (2), 2007, 223–237.

6

Unification:
An Important Way for Quick
Restoration of Relay Protection after
Intentional Destructive Impacts

One of the problems, which appears after intentional destructive impacts on relay protection systems, is quick restoration of its normal functioning and (in the first turn) quick restoration of DPR's workability. In order to achieve this restoration, a full set of functional spare modules of DPR needs to be available. When using dozen types of DPRs of different manufacturers at one energy-supplying company or even at one substation, it is necessary to have a corresponding amount of functional spare modules, which is unreal from practical standpoint.

The problem of changeability of DPR modules of different types is up-to-date not only in relation to the discussed problem but also under normal conditions of use.

6.1 Actual Situation and Problems in Unification of Construction of the Digital Protective Relays

Today, there are hundreds of MPD models of dozens of manufacturers in the market. Each type of DPR is built in a separate body, which may be totally different from that of any other type of DPR (even of the same brand) (Figure 6.1).

Usually separate DPRs are installed in relay cabinets: 3–5 units in each cabinet (Figure 6.2).

Historically, [1] there is a large number of noninterchangeable and incompatible DPR designs. This means that if a certain module of any DPR installed in the particular substation or power station fails, it can be replaced only by the same one produced by the same manufacturer. Thus, after you have spent a small fortune purchasing a DPR from one of the manufacturers, you fall into economic dependence on this manufacturer for the next 10–15 years, since after you have chosen one manufacturer, it no longer makes any difference if there are other manufacturers in the market, as you

FIGURE 6.1
Configuration of modern MPDs of different brands.

FIGURE 6.2
The method of DPR installation into cabinets in use today.

cannot use their products. And the only way to get out of this is to pay a small fortune one more time for the DPR from another manufacturer (and, thus, you switch from one bondage to another). And what does the manufacturer do with an absolute monopoly? Right: the manufacturer increases the price! The price of one spare DPR module can reach almost one-third or even a half of that of the entire DPR! As you have no other choice, you pay that price. And what happens after 8–10 years of the DPR's operation? Here's what: the manufacturer has already developed several new designs during that time, and it is unprofitable for the manufacturer to maintain the facilities producing spare modules for old relays, so the production has been stopped. What should a consumer do in this situation? Right: throw the old DPR into garbage, even if there is only one faulty module (printed circuit boards [PCBs] of modern DPRs are developed in such a way that they cannot be repaired), and fork up another sum of money for purchasing a new unit. Therefore, insufficient hardware reliability of DPR [2–5] results in a serious economic problem since a built-in self-diagnostics feature so much advertised by manufacturers does not help to reduce failures and breakdowns.

The popular way of increasing hardware's reliability as backup is also problematic due to the high cost of DPR and shortage of resources to reequip the basic set of protections apart from the backup set.

Another way of increasing the reliability of electronic equipment related to preventive replacement of limited life units, for example, power supply units with electrolytic capacitors, is hardly ever used in practice due to the same reasons.

Full incompatibility of DPR's software, sometimes even between different versions of the same program not to mention the software from different manufacturers, results in another problem hampering the use of DPR. The same energy company may use four to five types of DPRs, and service staff should learn all these essentially different programs resulting in serious problems due to the so-called human factor. And this happens in the background of ubiquitous sophistication of DPRs and their software. The following is the opinion of a leading Russian protection engineer concerning the relay protection devices from one of the world's leading manufacturers of DPRs [6]:

> Terminal Siprotec 7SJ642 (Siemens) has unreasonable technical and informational redundancy. User manual (C53000G1140C1476, 2005) declares "simple operation through integrated control board or PC with a systems program DIGSI," which is totally untrue. For example, you should enter nearly 500 parameters (settings), despite of inevitable changes to signal matrix, while each signal has its "properties" influencing the operation of the unit (printed-out DIGSI signals matrix takes about 100 pages of text in English). Since there is a need to compile terminal adjustment tasks, where all set-up protocols should be considered, the amount of documentation becomes huge. Big volumes of data need

to be entered making the setup process very difficult. Informational redundancy increases the probability of so called human factor errors. Technical redundancy requires only high-level specialists to be involved in work with the terminal. Available technical documentation includes thousands of pages, which often do not provide necessary data while containing errors.

The aforementioned Siemens product is not the only device with these kinds of problems. This is also true for any other brand. Unfortunately, this is a common trend today. The lack of simple and basic design and software standards significantly complicates acceptance and periodic testing of DPR [7].

The latest trend of increasing the number of protection functions in one module, "charging" excessive additional nonprotection functions, and implementation of so-called nondeterministic (not defined) logic additionally affects the reliability of relay protection and increases the unpredictability of DPR behavior under emergency conditions [8,9].

6.1.1 Is There Any Way to Solve These Problems?

The concept of DPR design we propose is based on the following basic principles:

1. The operating assemblies of the DPR should be distinctively well separated, and thus the chaotic layout of the assemblies on the printed boards [10] should be replaced with an ordered, standardized arrangement. For example, operating assemblies such as power source unit, input current, and voltage transformer with signal preprocessing units, digital input module, output relay module, central processor module, etc., should be arranged on separate printed boards of a standard size with universal connectors.

2. Separate relay protection devices for power plants and substations should be designed and sold as separate universal printed boards (modules), rather than a number of devices in enclosures of different sizes and shapes, to be used by a customer to configure required DPR. These boards (modules) should allow simple installation (by guide rails to connect to cross-board connector) in metal cabinets with separate sections and separate doors. Metal cabinets should be made under the technology aimed to protect the contents from external electromagnetic disturbances.

3. The DPR should perform only relay protection functions. The number of functions integrated in one module should be optimized according to cost and reliability and limited by standard.

4. The DPR's software should include a standard basic framework and set of different applications and libraries compatible with common basic framework.

5. All modules in the cabinet must be connected to two heavy-duty power sources, connected together as main and standby.

6. All these principles must be described in the new standard with a conventional name, such as "Principles of MPD Design. Basic Requirements."

6.1.2 Realization of the Proposed Concept

But is it possible to realize the proposed concept from a technical point of view? As mentioned earlier, the majority of DPRs available in today's market do not include a set of modules strictly separated by functions and design, and they remain a hodgepodge such as a CPU and switched power supply located on the same printed board [10]. We have analyzed a lot of different types of the most up-to-date DPRs of the world's leading brands and finally found devices, which ideally correspond to the preceding design requirements. These are MPDs series 900 of the well-known Chinese company Nari-Relays with universal modules used in protection devices of different types (Figure 6.3).

These turnkey modules do not require any preliminary preparation except of logic and setting configuring. Moreover, once assembled, the DPR does not require any adjustments. In order to assemble the DPR, you only need to install printed boards as shown in Figure 6.3 (the actual set includes power source board, which is not shown in the picture since it is not required as part of the set in our concept) into labeled rails of the enclosure (in our case, it is the section of the cabinet). Assembly of such complex protection as distance consisting of seven separate modules supplied in cartons and all assembling process doesn't take more than 10–15 min, after which you can begin entering the set points. It is obvious that an ordinary protection technician who does not have expertise in microprocessor technology won't have any problem with on-site assembling of protection relay consisting of such universal units.

Basically, nothing today prevents early implementation of the proposed concept in a given country. Even a small company can enter the DPR market offering consumers a new concept of cheap and reliable relay protection, equipped with backup block modules. It should start with purchasing sets of universal Nari-Relays modules (with different algorithms written in electrically erasable programmable read only memory [EEPROM] and different sets of input transformers) and mastering the production of cabinets.

6.1.3 What Are the Advantages of the Proposed DPR Development?

For a consumer:

- Significant reduction in DPR's cost value.
- DPRs may be composed of individual modules of different brands that best meet the needs of the operating organization in terms of optimal balance between quality and cost.

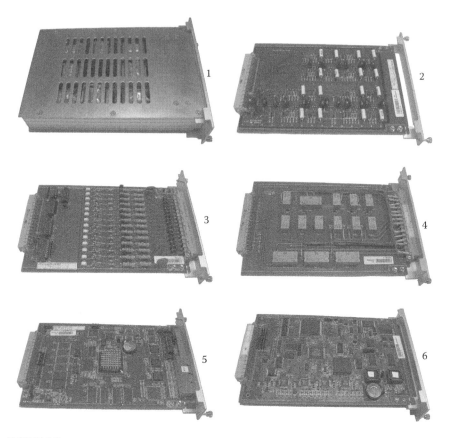

FIGURE 6.3
Set of universal functional units (220 × 145 mm) arranged on separate printed boards, which different Nari-Relays DPRs consist of PCS-931 (differential line protection), PSC-902 (distance protection), etc. (1) Input current and voltage transformers module, (2) low-pass filter (antialiasing filter), (3) digital input module, (4) output relay module, (5) optical communication module, (6) CPU module.

- Optimal spare parts and accessories (SPTA) set of DPR modules.
- Reducing the urgency of the problem of low reliability of DPR through quick and easy on-site replacement of failed low-cost modules with standby modules operated automatically, if the basic units are damaged, no need to repair faulty modules.
- Antimonopoly measures since you don't depend upon the manufacturer of DPR.
- Enforcing competition between producers thanks to new market players—small- and medium-size companies specializing in production of only certain types of modules rather than complete DPRs.
- Simplifying the testing of the DPR and reducing the influence of "human factor."

- Easier software, which allows selecting the most suitable and convenient application (interface) and painless replacement of DPR applications (interfaces).
- Acceleration of technological progress in the field of DPR without complicating the operation and any additional problems during upgrading devices to the next generation.
- Reduced the cost of updating the DPR, as the whole DPR will not have to be updated every 10–15 years as is the case today. You will only need to update some individual modules. Moreover, you will be able to update the CPU board more frequently than now, thus speeding up the technological progress in this area.

For a manufacturer:

- No need to produce outdated modules to old DPR installations.
- No lifetime free repairs.
- Significant increase in sales of individual modules.
- New emerging markets for basic framework and set of different applications and libraries compatible with common basic framework.
- Specialization in the production of individual and the most profitable types of modules.
- Small- and medium-size companies that do not have sufficient resources for the development and manufacture of complete DPRs may become the competent competitors.
- Competitive advantage for domestic manufacturers, who will be the first to start production of the DPRs as modules in a specific country, compared with foreigners.

The proposed direction of development could open the DPR market to new players: some of them would produce analog input modules equipped with current and voltage transformers, while the others CPU modules or software. The consumer could assemble the DPR from separate modules of different manufacturers, just as it is the case with PCs today, based on the cost and quality of these modules, as well as use the same software for all its DPRs. It would solve many of the issues raised earlier and significantly reduce the cost of relay protection. The latter could also allow installing two sets of identical protection types instead of one in order to improve reliability, while the other set would be used as backup, starting automatically upon receipt of the "watchdog" signal coming from the damaged core DPR. In addition, it would eliminate the necessity to use individual power sources for each DPR; rather, it would enable the use of one double high-capacity power source set of improved reliability for the entire cabinet instead. Finally, it would allow installing many service modules, capable of improving the DPR's reliability, in the same cabinet.

Thus, the relay protection maintenance would be simpler as the service staff would not need to read thick folios about different DPRs installed in the facility and study specific characteristics of the software of each DPR type. In addition to easier maintenance and time savings for installation and learning new types of protection, it would significantly reduce the percentage of errors caused by the so-called "human factor." Such design of DPR would solve the problems of testing complex DPR functions.

Of course, the share of DPRs in the market is significantly smaller than that of PCs with their 52 billion dollars in sales; it is still big enough for successful implementation of the proposed concept.

How should we start to implement this concept? We believe that the starting point should be the development of the aforementioned national standard by a wide range of professionals including protection engineers, relay operators, scientists, design engineers, and industry representatives, as well as a contract with the Nari-Relays, as a first step toward the realization of concept.

Unification of DPR modules will allow solving multiple problems under normal conditions of using and significantly increasing restoration of DPR after intentional destructive impacts.

6.2 Unifications in the Technical Specifications

The availability of unified (in terms of design) DPR modules, which were described in this chapter, before is not the only problem on board of quick restoration of DPR after intentional destructive impacts on relay protection systems. Besides compatible design, these unified modules should also be unified in terms of technical specifications based on common standards. Is there a universal technical specification accepted by all manufacturers and users of DPR? The answer is: No!

The following are basic problems of the DPR specification unifications:

- Current technical documentation refers to the standards of different groups: IEC 61000 (general standards for EMC), IEC 60255 (special standards for measuring relays and protection equipment), and ANSI/IEEE (Institute of Electrical and Electronics Engineers, United States), which are not fully compatible. Different manufacturers and consumers specify different standard groups in the technical and tender documents, thus significantly complicating the analysis of the relay characteristics and usage of the technical documentation.

- Some countries still rely on outdated national standards that do not correspond to the international standards.

- The technical documentation usually refers only to the standard numbers, without specifying the category (level) and acceptance criteria; therefore, it is not possible to determine the real parameters, since even in the same standard, the parameters may be two- or even threefold different from each other.

- Tender and project documentation contains wrong parameters and references to nonexistent categories, levels, and acceptance criteria of standards as a result of numerous subsequent copying of the parameters from the different specifications of the manufacturers.

- Some standards contain only references to other standards instead of specific numerical values.

- The potential consumer of the protection relays must know many standards and have a good understanding of complex issues out of their sphere of competence and professional activity (e.g., of EMC).

- The extreme complexity of the comparative evaluation of DPRs from different manufacturers stemming from reference to different groups of standards in the technical documentation or presenting technical data in different formats.

- The extreme complexity of determining the compliance of the various types of DPRs from different manufacturers with the requirements of the tender documents due to referring to different groups of standards.

As a result, the DPR technical documentation from manufacturers often contains the data that are incomprehensible or misleading to the consumer. Moreover, the manufacturers are not able to present clear, correct, and complete test condition data to the certification centers and testing laboratories in order to make necessary tests.

Conversely, the consumer is not able to properly formulate the technical requirements in tender documentation; compare the products from different manufacturers, whose technical parameters are shown in different forms and refer to different standards; or evaluate the compliance between the tender documentation requirements and characteristics of the proposed products since they have references to different standards. This problem is particularly severe if someone purchases DPRs from foreign manufacturers implementing IEC standards different from the national standards of the buyer.

This situation is a loss for both the manufacturers and the consumers of DPRs.

After analyzing dozens of DPR specifications from all the world's leading manufacturers and the set of standards for Russian Federation, IEC, and IEEE, I have developed a suggested Universal Basic Specification, which can be used by the manufacturers and the consumers of DPRs to address many of the problems mentioned earlier. This specification is based on the international standards IEC 60255 for DPR implemented in most countries of the world. Since the specification is based on the international standards, it can pull together the national and international manufacturers of DPRs and simplify the use of the DPRs from foreign manufacturers for the consumers.

Another important advantage of the international standards is that they are updated more frequently than national standards and reflect the current and actual data derived from the practice or research.

Recommended Universal Specifications on Digital Protective Relays

1. *Contacts*
   a. Contact Rating: IEEE St. C37.90
      Tripping Output
         Nominal Voltage: 250 V AC/DC
         Make: 30 A and carry for 0.2 s, inductive load ($L/R \leq 0.04$ s)
         Carry: 5 A continuous
         Break: DC 50 W resistive, 25 W inductive ($L/R \leq 0.04$ s)
         Break AC load: 1250 VA, $\cos \varphi = 0.7$
      Signaling Output
         Nominal voltage: 250 V AC/DC
         Make: 5 A and carry for 0.2 s, inductive load ($L/R \leq 0.04$ s)
         Carry: 3 A continuous
         Break: DC 30 W resistive, 15 W inductive ($L/R \leq 0.04$ s)
         Minimal contact load: 20 mA at 24 V AC/DC
         Break AC load: 500 VA, $\cos \varphi = 0.7$
      Contacts for Energizing Logic Inputs of Protective Relays
         Nominal voltage: 250 V AC/DC
         Minimal switching current: 1 mA at 250 V
   b. Durability
      Loaded contact: 10,000 operations minimum for signaling output; 1,000 operations minimum for tripping output
         Unloaded contacts: 100,000 operations minimum
2. *High-Voltage Withstand*
   a. Insulation Resistance
      IEC 60255-5
      >100 megaohm at 500 V, during 5 s minimum:
      i. Between independent circuits and all other circuits connected together with earth terminal
      ii. Across the open contacts of all the output relays

b. Dielectric Withstand on AC Voltage at Main Frequency

IEC 60255-5

   i. 2 kV rms, 50 Hz, 1 min between independent circuits and all other circuits connected together with earth terminal

   ii. 1 kV rms, 50 Hz, 1 min across the open contacts of the output relays

ANSI/IEEE St. C37.90

1.5 kV rms, 50 Hz, 1 min across open contacts of tripping relays

NOTE: For solid-state tripping relays with internal protective overvoltage elements, dielectric withstand voltage applied across output terminals of the relay in OFF condition is no more than 1.5 of rated voltage.

c. High-Voltage Impulse

IEC 60255-5, category IV

Three positive and three negative impulses at an interval of 5 s, 5 kV peak, 1.2/50 μs, 0.5 J between independent circuits and all other circuits connected together with earth terminal (except communication ports)

3. *Power Supply*

a. Operating Voltage Range

176–264 VDC, criteria acceptance A

b. AC Ripple on DC Supply

IEC 60255-11, criteria acceptance A

The protection relay will withstand without de-energizing, misoperation, and losses of data, 15% AC ripple (sinusoidal waveform) of rated DC value on the DC power supply.

c. Power Supply (Dips) Interruption

IEC 60255-11, criteria acceptance A

The protection relay must withstand such interruption in the auxiliary supply, under normal operating conditions, without de-energizing, misoperation, and losses of data:

   i. For DC power supply: voltage dips 100% during 10–1000 ms (according to manufacturer choice)

   ii. For AC power supply: voltage dips 100% during 0.5–25 cycles of power supply frequency (according to manufacturer choice)

d. Power-Up Time

Time to power up <60 s

4. *Electromagnetic Compatibility*
   a. High-Frequency Burst Disturbance

   IEC 60255-22-1, level 3, criteria acceptance A:

   i. 2.5 kV (peak), 1 MHz for common-mode test between independent circuits and all other circuits connected together with earth terminal (except communication ports)

   ii. 1 kV (peak), 1 MHz for differential (transverse) test across terminals of the same circuit (except communication port)

   iii. 1 kV (peak), 1 MHz for common-mode test between communication port and all other circuits connected together with earth terminal

   Test duration 2 s, 6–10 bursts per period of power supply frequency

   b. Fast Transient/Burst Immunity

   IEC 60255-22-4, level A, criteria acceptance A

   Common-mode test between independent circuits and case (earth):

   i. 4 kV (peak) for all circuits excluding communication port

   ii. 2 kV (peak) for communication ports

   Common-mode test between independent circuits and all other circuits connected together with earth terminal

   Repetition rate during burst 5 kHz; burst period 300 ms, burst duration 15 ms; test duration 1 min, each polarity

   c. Surge Immunity Test

   IEC 60255-22-5, level 3, criteria acceptance A:

   i. 2 kV (peak) for common-mode test between independent circuits and case (except communication port)

   ii. 1 kV (peak) for differential (transverse) test across terminals of the same circuit (except communication port)

   The pulse waveshape: 1.2/50 μs

   d. Immunity to Radiated Electromagnetic Energy

   IEC 60255-22-3, level 3, criteria acceptance A:

   The frequency range is swept from 80 to 1000 MHz with the signal 80% amplitude modulated with 1 kHz sine wave, field strength 10 V/m

   Addition spot frequencies: 80, 160, 450, and 900 MHz

e. Immunity to Conducted Radio-Frequency Interferences

IEC 60255-22-6, level 3, criteria acceptance A:

   i. 10 V (rms) for frequency 150 kHz to 80 MHz with amplitude modulated by frequency 1 kHz at 80% sine wave

   ii. 10 V (rms) for frequency 80–2700 MHz with pulse modulation

f. Power Frequency Immunity

IEC 60255-22-7, class A, criteria acceptance A:

   i. 300 V (rms) for common-mode test between independent circuits and all other circuits connected together with earth terminal (except communication port)

   ii. 150 V (rms) for differential (transverse) test across terminals of the same circuit (except communication port)

g. Power Frequency Magnetic Field Immunity

IEC 61000-4-8, level 5, criteria acceptance A:

   i. 100 A/m applied continuously

   ii. 1000 A/m applied during 3 s

h. Pulse Magnetic Field Immunity

IEC61000-4-9, class 5, criteria acceptance A

1000 A/m, waveform 8/20 μs, five positive and five negative pulses every 10 s

i. Damped oscillatory magnetic field immunity

IEC61000-4-10, class 5, criteria acceptance A

100 A/m for frequency 100 kHz and 1 MHz with a burst duration of 2 s, applied in all planes

j. Radiated Immunity from Digital Communications

ENV 61000-4-3, level 4, criteria acceptance A

Test field strength 30 V/m at frequency band 800–960 MHz and 1.4–2.0 GHz with the 80% amplitude-modulated signal, 1 kHz

k. Electrostatic Discharge Test (ESD)

IEC 60255-22-2, class 4, criteria acceptance A:

   i. 15 kV discharge in air to user interface, display, and exposed metalwork

IEC 60255-22-2, class 3, criteria acceptance A:

   i. 6 kV discharge in air to all communication ports

   ii. 8 kV point contact discharge to any part of the front of the protective relay

5. *Environmental Conditions*

(in accordance with concrete climatic zone or conditions)

a. Ambient Temperature Range

IEC 60255-6

Temperature:

i. Operating temperature range: −10°C to +55°C

ii. During storage: −10°C to +65°C

iii. During transportation: −40°C to +85°C

Humidity:

i. Relative humidity (annual average) 75%

ii. High-relative humidity during 30 days, 95%

Maximal altitude above sea level, 2000 m

6. *Mechanical Requirements*

a. Vibration

IEC 60255-21-1, class 2

i. Vibration response (energized): sinusoidal; frequency 60–150 Hz; acceleration 1*g*; sweep rate 1 octave/min; one cycle in three orthogonal directions. Criteria acceptance A

ii. Vibration withstand (de-energized): sinusoidal; frequency 60–150 Hz; constant acceleration 2*g*; 40 cycles in three orthogonal directions

iii. Vibration during transportation: 2*g* in each of three mutually perpendicular axes swept over range of 10–500 Hz for a total of six sweeps, 15 min each sweep, without structural damage or degradation of performance

b. Shock

IEC 60255-21-2, level 1:

i. Shock response (energized) semisinusoidal, 5*g* acceleration, duration 11 ms, each three shocks in both directions of three axes; criteria acceptance A

ii. Shock withstand (de-energized) 15*g* acceleration, duration 11 ms, each three shocks in both directions of three axes

c. Bump Test

IEC 60255-21-2, level 1

De-energized: 10*g*, 1000 bumps, 16 ms duration, on each direction of three axes

 d. Seismic

 IEC 60255-21-3, level 2, criteria acceptance A:

 i. *X*- and *Y*-axes: 3*g*, 11 mm, 1–50 Hz

 ii. Z-axis: 2*g*, 7.5 mm, 1–50 Hz

NOTE: "Criteria acceptance A"—normal performance of protection and control functions within specification limits, during and after the test, without de-energizing, misoperations, and losses of stored data or transmitted data.

It is safe to assume that the implementation of the Universal Basic Specification, supplemented with the specific parameters of the certain types of DPRs by all the DPR manufacturers and the consumers, will contribute to the addressing of various current problems listed earlier.

6.3 Unification in Evaluating Reliability of Digital Protective Relays

Reliability is defined as the property of an object to maintain over time, within a given range, the value of all parameters characteristic of its ability to perform required functions in predetermined modes of operation and conditions of use, maintenance, repair, storage, and transportation. As can be seen from this definition, reliability is a multidimensional property that may include, depending on the purpose of an object and the environment within which it is placed, fail-safe, durability, serviceability, and storage ability or some combination of any of these.

One of the key reliability indicators is the mean (operating) time between failures (MTBF), defined as total operation time (or the sum of the operational periods) of a *restorable* item divided by the number of observed failures within this time. That is to say, it is one of the reliability indicators of a *repairable* device or an engineering system characteristic of the average time (in hours) of device operation between failures (repairs).

6.3.1 Problems with Using MTBF to Evaluate Reliability of DPR

Manufacturer's technical manuals generally claim this period for DPR to be equivalent to 50–90 years. Does it mean that the time between two DPR failures is really 50–90 years? Despite the definition given to this term, common sense suggests that in real life, as opposed to virtual reality, it can't be so. As they say, more is the pity for common sense.

There are many variations of MTBF, for example, mean time between unit replacements (MTBUR) that is defined as arithmetic mean (average) time to failure (replacement) of a replaceable unit.

It is quite apparent that given modular design of DPR and nonservice-ability of multilayer PCBs with electronic surface-mounted devices (SMDs) being the basis of the state-of-the-art DPR, "replaceable units" might only mean integral modules (PCB), and DPR repair (restoration) may generally be carried out only by module (PCB) replacement. In this case, there is no practical difference between MTBF and MTBUR indicators, and the consumers will continue to stare bewilderedly at amusing many-digit numbers corresponding to 50–90 years and wonder what they can mean and how they correlate to between 15 and 18 years' real service life of DPR.

From where can such mind-blowing MTBF numbers be derived? Naturally, they may be obtained only theoretically through calculations. Briefly, these calculations appear as follows. Let us assume that 1000 units were subjected to test during the year. During the test, 10 units failed. Then MTBF will be equal to 1 year × (1000 units/10 units) = 100 years or, in round figures, 900,000 h. It is this many-digit number that the consumer will see in technical manual or information sheet in respect of DPR.

But why then don't DPRs last this long if the calculations suggest they should? There may be dozens of reasons for this. First, testing during one (or even more) year does not allow consistent failure results since failure rate varies substantially over time and application of a constant failure rate (as is the preceding case) does not ensure consistent results. In actual practice, failure rate over time is constant only in single region and is described by the Weibull–Gnedenko function (Figure 6.4).

In the case of a variable failure rate (i.e., when $\lambda \neq$ const), the preceding example of MTBF calculation is irrelevant, and it should be calculated using other, far more complicated formulae.

Second, in actual practice, many manufacturers, instead of pilot testing large quantities of their units in field operating conditions (which is both costly and time-consuming effort), carry out theoretical calculation of MTBF based merely on failure rate data for basic electronic components contained in DPR and on

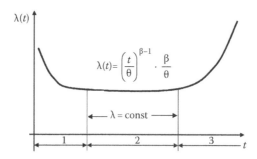

FIGURE 6.4
Failure rate–time relationship: (1) running-in period (early failures); (2) normal operating period (random failures); (3) deterioration period (wear-out failures); t, unit in-service time; θ, scale parameter; β, shape parameter. In running-in period, $\beta < 1$; in normal operating period, $\beta = 1$; and in deterioration period, $\beta > 1$.

their number in DPR. This calculation appears as follows. A device is comprised of, say, 10 components having a failure rate of $10^{-7}\,h^{-1}$ each. Then the device failure rate on the whole would be $10 \times 10^{-7}\,h^{-1} = 10^{-6}\,h^{-1}$, and the time to failure would amount to $10^6\,h$ = 1 million h. And this is where many uncertainties arise that cannot be foreseen in any calculations. High-quality electronic components themselves supplied by a renowned and trustworthy manufacturer may safely operate as a part of equipment for dozens of years and have rather low failure rates. But it is only the case in particular operating conditions for which these parts are. It is for these conditions that failure rates are referenced in parts' reference sources. It is these failure rates that are assumed in calculations carried out by DPR manufacturers. But what's the real state of affairs?

Example 1. Electrolytic capacitors intended for DPR switched-mode power supplies. Even high-quality general-purpose industrial grade electrolytic capacitors produced by well-known Japanese manufacturers fail fairly soon when affected by high-frequency currents flowing through them in switched-mode power supplies (see Figure 6.5). Leaking electrolyte results in the substantial damage of many other circuit components as well and even conductors and via interconnections in a PCB.

Power supply failure for this reason occurs after some 12–15 years of operation in DPRs of various types produced by different manufacturers. The problem is brought about by the wrong choice of electrolytic capacitor types by DPR manufacturers, lack of technology used to protect electrolytic capacitors from high-frequency currents in DPR circuits resulting

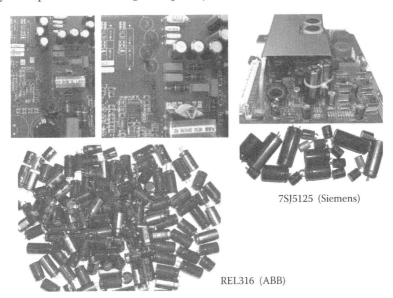

7SJ5125 (Siemens)

REL316 (ABB)

FIGURE 6.5
Faulty DPR switched-mode power supplies of different types with damaged electrolytic capacitors. Left figure shows taints in PCBs from electrolyte leakage.

FIGURE 6.6
A part of DPR logic input module of REL316 type with damaged capacitors C and failed optocouplers Opt.

in electrolyte heating, and increase of its chemical activity. Has this problem with electrolytic capacitors been taken into account when calculating MTBF? Obviously not!

Example 2. Disk ceramic capacitors encased in a molded plastic shell (see Figure 6.6). In DPRs used or operated in subtropical climates with high air humidity whose capacitors often lead to DPR failure due to the conduction path between the capacitor plates resulting from the migration of silver ions from one ceramic disk surface to another induced by applied voltage in high humidity environment when the capacitor sealing is not really perfect. As a result, ceramic capacitors generally known to be highly reliable components with long life factors result in multiple DPR failures after some 15 years in operation. Has this problem been taken into account when calculating MTBF? Obviously not!

Example 3. Transistor optocouplers abounding in input module circuits of any DPR (see Figure 6.6). The specific feature of an optocouplers is the gradual decrease of current transfer ratios (CTRs) caused by the degradation of optical plastic (decrease in transparency) used to connect light-emitting and light-detecting components of an optocouplers. Consequently, if the operation mode for optocouplers inside a DPR has been chosen in the initial section of characteristic (to limit the power dissipated by logic input circuits),

FIGURE 6.7
EEPROM components manufactured in 1996 that failed after 15 years of operation against the backdrop of extracts from technical manuals that guarantee retention of data recorded therein for 100 years.

then after 13–16 years of DPR operation, epidemic failures of their logic inputs will occur. Has this problem been taken into account when calculating MTBF? It's black and white.

Example 4. In technical manuals for such essential components of any microprocessor units as EEPROM, inherent data retention is claimed greater than 100 years (see Figure 6.7). Yet in actual practice, they have started to "clear" the data recorded therein as early as after 15 years of operation inside a DPR. Has this effect been taken into account when calculating MTBF?

An article by the employees of a DPR [11] manufacturer claims that their relays have an MTBF of 74 years and every single failure was detected at the time of operation by the in-house embedded DPR self-diagnostic system. Let us beg leave to doubt the credibility of such claims since no in-house embedded DPR self-diagnostic system is able to detect capacitor electrolyte leakage, or degradation of optocoupler's transfer ratios, or higher rates of self-discharge of flash memory components, or the problems with the control element called watchdog. As a result, we have a burst-type DPR failure flow occurring after 15–18 years of operation, while manufacturers claim an MTBF to be 50–90 years.

Interestingly, these kinds of problems have never occurred with electromechanical protective relays that have served hand and foot (and in fact are still in service) for many dozens years. The examples include inverse time relays of RI types that were manufactured about 100 years ago by Allmanna Svenska Elektriska Aktiebolaget (ASEA) (in English spelling, General Swedish Electrical Limited Company) in a Swedish city of Vasteras. In the upper left corner of the relay, you can see a swastika bearing the letters A, S, E, and A— logo (trademark) of ASEA Company (see Figure 6.8) placed on these relays up to 1933 when this symbol was assumed by German Nazi. Quite a few such relays were in service in power sector of the former Soviet Union (with the swastika carefully defaced), and they are familiar to the old generation of protection engineers. Until recently, these relays could still be found operating on the sites including in the territory of Russia, and it was not because they could no longer fulfill their functions that they were replaced, but rather, it was nothing but a shame to keep using a hundred-year-old relays any more.

Over the past few years, the professional community has developed an awareness of the fact that DPRs are less reliable than electromechanical relays. The solution of this problem is generally thought to be DPR redundancy.

FIGURE 6.8
Electromechanical relay RI (ASEA) manufactured about a hundred years ago that retained its operability to this day.

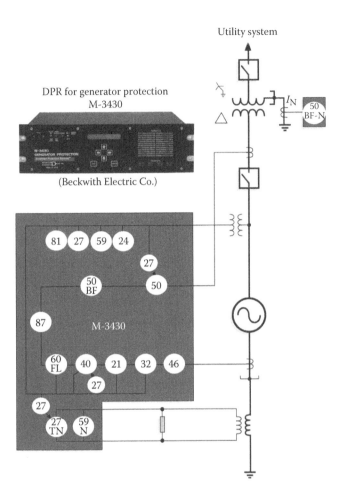

FIGURE 6.9
M-3430 multifunction DPR. The numbers shown in white circles designate standard relay protection functions under ANSI classification.

The problem is becoming ever more relevant with the number of functions being performed by a single DPR terminal. In multifunction relays that "put all the eggs in one basket" (see Figure 6.9), a failure or malfunction of only one of these "eggs" may result in the disconnection of a generating unit, thus causing great damage. For this reason, if multifunction DPRs are used to protect critical objects, manufacturers themselves advise [11] (notwithstanding the MTBF values claimed by them to be equal to many dozens of years!) to use double DPR sets (see Figure 6.10). This way, the calculated value of the MTBF per such a double set has been obtained in Ref. [12] to be equal to 500 years!

Here, we face some more questions. First, what are such absolutely fantastic MTBF values having nothing to do with reality for, and what are they worth?

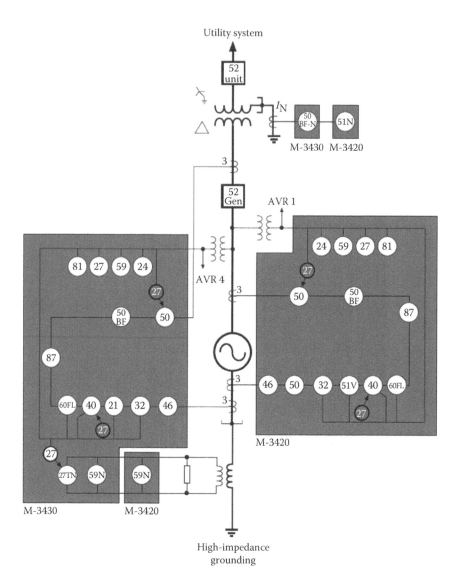

FIGURE 6.10
Double (redundant) generation unit protection set for more reliable protection.

Second, massive accidents in power grids may be caused by both the failure to shut off the sections running in the emergency mode and the false tripping of healthy grid sections (generating units, loaded lines) with load swing to other generating units and loaded lines (this scenario was pursued in one of the largest accidents in the United States). It means that DPRs are subject to two rather than one faulty states: both failure to operate and false operation. Herein, practical use of two identical sets, live and standby, is not all that simple since it is unclear how in this case you should connect DPR output contacts actuating the circuit breakers, in logic AND circuit or in logic OR circuit.

Any one connection option reduces the probability of one DPR faulty state while accordingly enhances the probability of the other. That is, the use of two identical DPR sets is apparently inadequate to enhance the reliability of relay protection for critical objects, and it is wise to use three sets with output signal majorization based on "two out of the three" principle.

One more problem relating to the MTBF application may occur in the near future. The market entry by versatile functional modules [13,14] sold and acquired as stand-alone products that are used to construct DPRs (as is the case with PC desktops today) moves these individual PCB modules from the "replaceable component part" category to the category of "stand-alone nonrestorable part," items that are highly versatile and have different reliability values. It is obvious not only that in this case, reliability values will have to be calculated on a per module basis but also that the MTBF rate cannot be applied to them collectively since they are nonrestorable items.

One more doubt as to the application of MTBF to DPR is that even single failure damages may be very high indeed; hence, a substantial time span between the first failure and the second failure (high MTBF rate) will be of little use.

6.3.2 New Criterion for DPR Reliability Evaluation

Considering that MTBF indicator has completely defamed itself by great values having nothing to do with reality and giving no actual information on DPR reliability and by its obvious limitations, application of MTBF for DPR reliability measurement should be dropped.

A new DPR reliability indicator is recommended [15]: *gamma-percentage operating time to failure (operating life, operating service)*, that is, the time during which an item failure shall not occur with a particular probability expressed as a percentage. For example, 95% operating time to failure within at least 5 years means that during 5 years of operation, failed devices shall make up a maximum of 5% of all devices in service. Besides, this value shall be specified for both DPR as a unit and separate PCB functional modules of which it is comprised. With such an intuitive and straightforward indicator, the consumer could trace the number of failed DPRs (or separate modules from which it is built) during a particular period of time and make claims against the manufacturer if within the observed time many more DPRs failed than what is guaranteed by the manufacturer. With such an indicator, it is much easier for the consumer to be guided in the future market of versatile modules [14] to choose the most cost-effective alternative.

Besides, manufacturers shall be required to specify, in both technical and tender documentations, average service life for individual modules and include guidelines on the frequency of preventive replacement of these modules to maintain high reliability of relay protection. Such periods may amount to, for example, 8–10 years for power supplies, 12 years for logic input modules, 15 years for central processor units, and 17 years for analog input modules. These data shall be known to manufacturers who keep a close watch on product failure and damage statistics respecting the code of good practice. The question of who shall bear the costs of such preventive module replacement shall be decided by agreement between a manufacturer and a consumer. For example, a manufacturer might guarantee nonrecurring (possibly partial, e.g., covering power supplies only) preventive module replacement, while any further replacements shall be carried out at the expense of a consumer. Large-scale preventive maintenance has already been carried out upon the author's recommendations (although it is limited to electrolytic capacitors contained in healthy DPR power supplies, type REL/REC/RET, series 316, manufactured by ABB using technology that still allows such replacement) in an power company operating many DPR of this series. The question of commencement of capacitor's preventive replacement in DPR power supplies produced by Siemens after 10 years in operation is now pending.

The use of suggested criterion for measuring DPR reliability and of additional reliability data discussed earlier will make it possible to change the nature of relationships between DPR consumers and manufacturers and to enhance reliability of relay protection. Practical implementation here depends on the consumer who is to specify appropriate requirements to DPR reliability in tender documentation along with basic technical requirements [16] since soon changes of regulatory documents are nothing to hope for.

References

1. Gurevich V. I. Problems of microprocessor protective relays: Who is guilty and what to do? *Energo-info*, 10, 2009, 63–69.
2. Gurevich V. Reliability of microprocessor-based relay protection devices: Myths and reality. *Serbian Journal of Electrical Engineering*, 6 (1), 2009, 167–186.
3. Gurevich V. I. Some performance and reliability estimations for microprocessor based protection devices. *Electric Power's News*, 5, 2009, 29–32.
4. Gurevich V. Reliability of microprocessor-based protective devices—Revisited. *Journal of Electrical Engineering*, 60 (5), 2009, 295–300.
5. Gurevich V. Problems of microprocessor protective devices: Specialist's opinions, unsolved problems and publications. http://digital-relay-problems.tripod.com/.
6. Belyaev A., Shirokov V., Yemelianzev A. Digital terminals of the relay protection. Experience in adaptation to Russian conditions. *Electrical Engineering News*, 1, 2007, 38–40.
7. Gurevich V. Tests of microprocessor-based relay protection devices: Problems and solutions. *Serbian Journal of Electrical Engineering*, 6 (2), 2009, 333–341.

8. Gurevich V. I. Sensational "discovery" in relay protection. *Power and Industry of Russia*, 23–24, 2009.

9. Gurevich V. Sophistication of relay protection: Good intentions or the road to hell? *Energize*, Jan/Feb, 2010, 44–46.

10. Gurevich V. Microprocessor protective relays: How they constructed? *Electrical Market*, 4, 2009, 46–49; 5, 2009, 46–50; 6, 2009, 46–50; 1, 2010, 2, 3.

11. Mozina C. J., Yalla M. V. V. S. Design, manufacturing and application of multifunction digital relays for generator protection. Canadian Electrical Association, Montreal, Quebec, Canada, 1996.

12. Ward S. Improving reliability for power system protection, relay protection and substation automation of modern power systems (RFL Electronics Inc., USA), Cheboksary, Russia, September 9–13, 2007.

13. Gurevich V. I. Relay protection: Thinking about future. *Electrical Networks and Systems*, 1, 2011, 73–80.

14. Gurevich V. I. The new concept of digital protective relays design. *Components and Technologies*, 6, 2010, 12–15.

15. Gurevich V. I. Problems with evaluations of the reliability of relay protection. *Electrichestvo*, 2, 2011, 28–31.

16. Gurevich V. I. Problems for standardization of the digital protective relays. *Components and Technologies*, 1, 2012, 6–9.

Epilogue

The main mistake people make is that they fear current problems more than future ones.

—Carl von Clausewitz

Some protection engineers think that their professional responsibilities are limited to the operation of relay protection (RP). They claim that they are absolutely not interested in the problem of protection of RP from intentional remote destructive impact (IRDI). They think that these issues should be a responsibility of "relevant authorities." These specialists claim that they have enough problems with RP; that's why they are totally against of any additional protection measures, which can make their difficult life even more difficult.

Many specialists are very excited about any innovations and new trends in the field of RP, regardless of where these innovations can take us, whether it is a "proactive relay protection" or RP based on "artificial intellect" or a "digital substation" or a "smart grid"… what's important here is to harvest enough money for development and adoption. Why not? Why not make some extra money? Money has no scent! Moreover, since these airy-fairy schemes are well funded and there are even governmental programs to support this trend, why not invent some other "intellectual" toys in RP? The important thing is to provide beautiful substantiation and use a lot of scientific words in descriptions.

There are specialists convinced that if processing capabilities of modern microprocessors used in digital protective relay (DPR) not only enable protection of RP but also do a lot of other things, then why not use these capabilities? Why not connect DPR to a number of different sensors and not use it for the monitoring of electrical equipment status? Why not use a DPR as an information and measurement unit or as a substation controller of automatic control system (ACS) in addition to its basic functions? In other words, based on the principle "the more the better" when "possible," it is perceived only as "necessary."

Some authorities responsible for making strategic decisions in the field of RP development are following the principle: if any developed country started moving in a specific direction, we should not lag behind; we need to catch up with them and outpace them! This policy reminded me of a joke, which I heard from an Englishman—a good specialist in the field of RP:

> There lived a large tribe of Indians near one of American cities. They were unhappy with their weather foreteller, since he was often mistaken. The tribal leader decided to call a weather forecasting center, since weather is an important factor in the life of Indians. The center replied that there is no exact forecast yet, but they suppose that the forthcoming winter will be cold. Having received this answer the tribe started collecting firewood. Just in case. Later, the tribal leader decided to update the forecast and called the forecasting center again. The answer was immediate: "Yes, of course, this winter will be extremely cold! Now we are sure! Look how Indians are collecting firewood for the winter, they know what they are doing!"

Many specialists speculate like this: we want to move forward toward technological advances, and your job is to protect us from all those IRDI! But it is high time to understand that under existing trends of RP development toward "digital substations," "smart grid," and protection relays with artificial intellect, where DPR not only is used as a protection relay but also exchanges information via multiple communication channels, it is impossible to ensure reliable protection of DPR from existing dangers.

Many advocates of DPR may think that I am opposed to everything new and advanced. One of the Russian renowned authorities in the field of RP, who occupies a top management position, wrote in his review to my article, where I showed the danger of modern trends of RP development that "the author tries to freeze technical advancement in Russia." These types of technical advancement advocates distort my attitude to technical advancement either mistakenly or on purpose. I do not call for neglecting innovations in the field of diagnostics of electrical equipment and information-computer systems, but rather separate them from RP. *I deeply believe that in order to ensure efficient protection of DPR from IRDI and increase reliability of its operation under normal operating modes, it is necessary to use DPR to perform only tasks of relay protection.* I think that after the period of excitement related to new opportunities has emerged with using microprocessor equipment in RP, it is now time to have a sober glance at these things and reevaluate the situation.

However, it is sad to admit that the majority of specialists and authorities responsible for solutions of today's problems have little interest in the potential problems and dangers of existing trends in RP development. This is all clear because none of them will be responsible for the collapse of the power supply system in case of IRDI due to lack of instructions or standard operation procedures (SOPs) on this issue. On the other hand, there are honors and awards for the introduction of new machines and adoption of new technologies.

Probably, the developers of electromagnetic and cybernetic weapons give today's tendencies in protective relaying an ovation, like the snake that is satisfyingly watching a frog trying to jump into its mouth. We can only trust that the wisdom of experts and officials will take precedence over the short-term mercantile interests and hope that as expressed in the well-known proverb "you don't have to wait for the rain to be falling to make a raincoat."

Index